ADVANCES IN SOIL AND WATER CONSERVATION

Edited by

F. J. Pierce
W.W. Frye

CRC Press is an imprint of the
Taylor & Francis Group, an informa business

CRC Press
Taylor & Francis Group
6000 Broken Sound Parkway NW, Suite 300
Boca Raton, FL 33487-2742

First issued in paperback 2019

ISBN-13: 978-0-367-44790-8 (pbk)
ISBN-13: 978-1-57504-083-7 (hbk)

Visit the Taylor & Francis Web site at
http://www.taylorandfrancis.com

and the CRC Press Web site at
http://www.crcpress.com

Cataloguing-in-Publication Data

Advances in soil and water conservation / edited by F. J. Pierce, W. W. Frye.
p. cm.
Includes bibliographical references and index.
ISBN 1-57504-083-2
1. Soil conservation. 2. Water conservation. I. Pierce, F. J. (Francis J.)
II. Frye, Wilbur W.
S623.S5715 1998
631.4′5—dc21 97-51557
CIP

Acknowledgments

We are grateful for the financial support of the Soil Science Society of America in sponsoring this symposium. A special thanks to Dr. David E. Kissel, President of the Soil Science Society of America, for his support, his participation in the symposium, and his presentation of a plaque to Dr. Gary Steinhardt, President of the Soil and Water Conservation Society, in honor of the Society's 50th anniversary. We also extend thanks to Skip DeWall of Ann Arbor Press, Inc., for his willingness to work with the editors to publish this book and to establish a forum for future discussion on this important topic in the Advances in Soil and Water Conservation series.

About the Editors

Francis J. Pierce is a professor of Soil Science in the Department of Crop and Soil Sciences at Michigan State University where he conducts research on soil management and precision agriculture and teaches a course *Soil Management and Environmental Impacts*. He received his B.S. degree in Geology from the State University of New York at Brockport in 1976 and his M.S. and Ph.D. degrees in Soil Science from the University of Minnesota in 1980 and 1984, respectively. Dr. Pierce has focused his research on conservation tillage systems and their impact on crop production and soil and water quality. He has worked to develop the concept of soil quality and methods to assess changes in soil quality in relation to land use and has worked to develop and evaluate site-specific management (SSM) for agriculture. He is a popular speaker on the subject of SSM and writes a column on SSM *Precise Advice* for the ag/INNOVATOR magazine. Dr. Pierce served as a member of the USDA-NRCS Blue Ribbon Panel from 1995-1996 whose report *Data Rich, Information Poor* has been distributed widely within NRCS. He has also served on the Board of Agriculture, National Research Council Committee on Conservation Needs and Opportunities in 1986 and was co-recipient of the USDA Distinguished Service Award in 1992 for his contributions to the NLEAP model to predict management effects on nitrate leaching. He has served as chairman of the Soil & Water Management & Conservation Division (S6) of the Soil Society of America. Dr. Pierce has consulted with USDA-NRCS on their role in precision agriculture and in establishing criteria for assessing soil quality in the U.S. He serves on the editorial boards of the *Journal of Sustainable Agriculture* and *Precision Agriculture Journal*. He has edited two other books, one by SWCS *Soil Management for Sustainability* and another by ASA *The State of Site-Specific Management for Agriculture*.

Wilbur W. Frye is Professor of Agronomy and Director of Regulatory Services at the University of Kentucky. He received his B.S. and M.S. degrees from the University of Tennessee and his Ph.D. degree from Virginia Polytechnic Institute and State University. Dr. Frye's career includes teaching, research, and administration at Tennessee Technological University and the University of Kentucky. At the University of Kentucky since 1974, he has taught soil science, and he and his colleagues have focused on multidisciplinary research on conservation tillage that involves the use of legume cover crops, soil and water conservation, soil properties, soil nitrogen transformations, nitrogen fertilizer efficiency, and energy conservation. He has authored or co-authored 148 publications, including 20 book chapters. Dr. Frye has served as associate editor of the *Soil Science Society of America Journal*; president of the Southern Branch American Society of Agronomy; member of the International Board of Directors of the Soil and Water Conservation Society; co-editor of ISTRO-INFO; member of the Committee on Conservation Needs and Opportunities, Board of Agriculture, National Research Council; and member of the Board of Directors of the Association of American Feed Control Officials. Currently, he is on the Board of Directors of the Council for Agricultural Science and Technology. His teaching honors include the Resident Education Award from the American Society of Agronomy, Soil Science Education Award from the Soil Science Society of America, Great Teacher Award from the University of Kentucky Alumni Association, Gamma Sigma Delta Master Teacher Award, Daryl E. Snyder Award, and a Danforth Associate. He is a Fellow of ASA, SSSA, and SWCS.

List of Contributors

R.R. Allmaras
Soil Scientist
USDA-Agricultural Research Service
Dept. of Soil, Water, and Climate
University of Minnesota
St. Paul, MN 55108

J.D. Bilbro
Big Spring, TX 79720-7020

R.L. Blevins
Dept. of Agronomy
University of Kentucky
Lexington, KY 40546

O.C. Burnside
Professor, Dept. of
Agronomy and Plant Genetics
University of Minnesota,
St. Paul, MN 55108

Pierre R. Crosson
Resources for the Future
1616 P Street N.W.
Washington, D.C. 20036

Richard Cruse
Professor, Department of Agronomy
3212 Agronomy Bldg.
Iowa State University
Ames, IA 50011-1010

T.C. Daniel
Dept. of Agronomy
115 Plant Science Building
University of Arkansas, Fayetteville,
AR 72701

J.W. Doran
Soil and Water Conservation Research
Unit
USDA-ARS
Lincoln, NE 68583

William M. Edwards
USDA-ARS
North Appalachian Experimental
Watershed
P. O. Box 478
Coshocton, OH 43812

W.W. Frye
Div. of Regulatory Services
University of Kentucky
Lexington, KY 40546

Donald W. Fryrear
Agricultural Engineer
USDA-ARS
Big Spring, TX 79721-0909

Dennis Keeney
Professor, Department of Agronomy
Director of the Leopold Center for
Sustainable Agriculture and of the
Iowa State Water Resources
Research Institute
209 Curtis Hall
Iowa State University
Ames, IA 50011-1050

W.D. Kemper
National Program Leader in Soil
Management (retired)
USDA-ARS
Building 005, BARC-WEST
Beltsville, MD 20705

Peter F. Korsching
Department of Sociology
Iowa State University
Ames, IA 50011

John M. Laflen
Agricultural Engineer
USDA-Agricultural Research Service
National Soil Tilth Laboratory
2150 Pammel Drive
Ames IA 50011

Preface

Beginning with the Dust Bowl of the 1930s, dedicated people like Hugh Hammond Bennett devoted their life's work to solving the soil erosion problems plaguing our nation. Out of the deep commitment of the early advocates for soil conservation, at times their rhetoric approaching evangelism, research in soil erosion and soil and water conservation emerged. As the scientific understanding of soil erosion and its control grew, so grew the ranks of professional soil conservationists. The need to share scientific knowledge among conservationists and to transfer conservation principles and practices to land owners gave birth to the Soil Conservation Society of America in 1946, today known as the Soil and Water Conservation Society (SWCS).

The year 1996 marked the 50th anniversary of the Soil and Water Conservation Society. In honor of this special anniversary, the Soil Science Society of America's Soil and Water Management and Conservation Division (S-6) sponsored a symposium at the SWCS 1995 annual meeting held in August in Des Moines, IA. The symposium was entitled **"Fifty Years of Soil and Water Conservation Research: Past Accomplishments and Future Challenges"**. The theme of the symposium was to highlight the contributions of research as the fundamental basis for soil and water conservation in the United States. This book completes the tribute to the Soil and Water Conservation Society and to the importance of research to soil and water conservation on the land.

This book is an in-depth, scholarly treatment of the most important developments and influences shaping soil and water conservation in the last 50 years. The book addresses not only the technological developments relating to an understanding of erosion processes and to methods for their control but also the policy and social forces that shaped the research agenda through that period and how this history may influence soil and water conservation in the future. Topics covered in this book include key governmental agencies and programs, research on processes of soil and water degradation, development of control practices and soil quality enhancement with emphasis on conservation tillage, the connection between soil and water conservation and sustainable agriculture, and the ways in which technology and social influences have and will shape soil and water conservation in this country. The authors dwell sufficiently on the past to establish for the reader a historical perspective on these major topics, but the present and future are the major focus of this book. The concluding chapter explores what the future of soil and water conservation might be in the next 50 years.

The historical foundation, the focus on key developments in soil and water conservation, the depth of treatment and thorough documentation, and the orientation to the future will make this book interesting and informative to all interested in soil and water management and conservation. As researchers and teachers on this subject, we understand and appreciate the vital importance of soil and water resources and the dependency of civilization on their quality.

We also know and appreciate the splendid qualifications of the authors. We thank them for their efforts. Most of all, we commend them on their superb contributions to what we hope is the first book in a series on *Advances in Soil and Water Conservation.*

Wilbur W. Frye and Francis J. Pierce

Introduction

W.E. Larson

The twenty-first century approaches, with increasing demands on the nation's and world's natural resources. In spite of a long history of soil and water conservation in this country, significant challenges to a sustainable natural resource base remain our legacy to the next century. World population continues to increase at an alarming rate requiring greater amounts of food and fiber products. Already limited land resources are further stressed by degradation processes like soil erosion, by extensive urban and industrial development, and by an increasingly diverse array of land uses, including those concerned with wildlife preservation, biological diversity, recreation, and wetlands. The adequacy of water resources, both in terms of quantity and quality, for human consumption or public and commercial uses is increasingly of concern to more areas of the U.S. and the world. The marking of the turn of the century is an opportune time to take stock of our progress in natural resource management and to assess the state of knowledge needed for the management and conservation of our soil and water resources.

Historically, research and government programs in soil and water use have been closely linked in the United States. Jefferson, Washington, and others warned of the degradation of soil resources occurring under the farming practices of their time and suggested that improved conservation and management must be implemented (Betts, 1953). Unfortunately, it was not until the passage of the Land Grant College Act in 1862 that major federal legislation was enacted in support of agricultural research. Subsequent to that landmark legislation, many agricultural experiment stations were founded in connection with the Land Grant Colleges by the end of the nineteenth century. More experiment stations followed in the early twentieth century to help guide the settlement of the West. It took the drought-induced erosion of the 1930s to spur the establishment of the Soil Conservation Service (now the Natural Resources Conservation Service, NRCS), which today still provides technical assistance to land managers.

The research focus in soil and water conservation during World War II shifted to food production in support of the war effort. Following the war, plentiful and cheap supplies of fertilizer and vastly improved plant genetic materials appeared, rapidly increasing our agricultural production capacity. During this 'green revolution,' protection of natural resources was neglected, at least for the moment. Neglect turned to concern with a strong environmental movement in the 1970s following increasing evidence of soil degradation and air and water contamination. Soil erosion control and water contamination from chemicals and sediment attained heavy emphasis in both research and public policy developments. These issues and debates on land uses and their links to soil, water, and air quality are still very much with us today.

During the twentieth century, soil and water use and conservation in the U.S. changed materially. Total cultivated land area decreased as crop yields increased dramatically in response to increased use of fertilizers and pesticides and improved genetics. Tillage practices and crop sequences changed significantly, while farm units became larger. Agricultural water use increased, as did the number of irrigated acres, with water quality and water availability declining in some areas. Soil erosion declined in many areas but remains unacceptably high in others. All of these changes have brought new questions as to how the nation can use its natural resources for the benefit of its citizens and the peoples of the world, and still preserve their quality for use by the people of the twenty-first century, and beyond.

Approaching the Twenty-First Century

Research contributions in soil and water conservation during the past century have been most impressive indeed. Progress in our understanding of erosion processes has made erosion prediction a major tool in identifying and targeting treatments to erodible land. Erosion control and water conservation has greatly improved through the development of cost-effective conservation tillage practices. The importance of wetlands is recognized and progress is being made in their preservation and their utilization in enhancing water quality and biological diversity. Problems with use of agrichemicals have been identified and improved management systems have been, and are, being developed. Overall, issues regarding the sustainability of agricultural production systems are being discussed and researched. These, and many other items, are openly discussed and solutions are being found. But the job is not finished, as soil and water conservation remains a critical and unresolved issue worldwide.

Current technologies and policies for the sustainable use and protection of our natural resources must be implemented promptly to meet the emerging food and fiber production needs of a growing world population without degradation of our natural resources. Logically, high levels of production should be concentrated on our best lands that are least sensitive to resource and environment degradation. Lands that are highly productive but with resource degradation hazards should be used with adequate safeguards. Lands that are not productive for agriculture should be designated for other uses such as for wetlands, wildlife, recreation, and other less intensive uses. Proper designation of lands for which they are best suited, and moving those lands into that use, may be one of the most fruitful efforts in the coming decades.

Soil erosion remains a major cost to the nation both from on-site and off-site damages. While erosion amounts have been reduced by about one-third since 1977 (estimates from the National Resources Inventory, NRI), erosion still is a major threat to continued high levels of crop production. Pierce et al. (1984), using a simple model and input from the 1977 NRI, concluded that the average potential productive capacity of soil would be reduced from 2 to 8% in the 15 Major Land Resources of the Cornbelt if erosion continued for another 100 years at the 1977 rate. The loss in potential productivity, however, on some of the steeper soils would exceed 20%, and eliminate them from economical row crop use. A number of investigators have reported that the off-site damages from erosion exceed that of on-site damages (Colacicco et al., 1989). Research on the development and implementation of improved tillage and residue management systems for erosion control and water conservation have been a major success story in the past several decades, and needs to continue if progress on erosion control and water quality is to proceed.

As crop yields are pushed ever higher, the quality and quantity of water available to plants will become the most limiting crop yield factor on much of our land. Increased competition for water from industrial and domestic needs may cut into water available for agriculture, and increase costs. Major advances are being made both in the efficiency of water use under irrigation and in rain-fed agriculture. Improved residue management has shown big gains in water storage efficiency in the semiarid regions. Research efforts must be intensified in water use eficiency for both irrigated and rain-fed conditions.

Along with water quantity, preservation of water quality is a major current issue. This century has shown a marked increase in the concentration of nitrates in the Lower Mississippi River and its tributaries, as well as in other surface waters, a consequence attributable to the use of nitrogen (N) fertilizers and excessive manure applications (Turner and Rabalais, 1991). Nutrient losses to surface waters have concerned human health authorities and natural resource scientists. Of wider concern is nitrates in groundwaters, particularly those used for human consumption, although concentrations are rarely at levels exceeding health standards (Holden et al., 1992; Power and Schepers, 1989).

Phosphorus (P) is a primary cause of eutrophication of surface waters since above a critical value, and when other nutrients are not limiting, aquatic vegetation will flourish and fish populations will decline. The natural levels of available P content of soils are generally low but levels often vary considerably within fields due to fertilization and manure applications and uneven plant extraction of P. Because the P content of runoff waters and eroded sediment is related to the available P content of the surface soil, it is important to only apply P fertilizer on those soils needing it for optimum plant growth. If P fertilizer is applied, it should be placed into the soil and not applied on the surface. Improved P management should be a national imperative.

Herbicides have been widely found in both surface and groundwaters within the last decade and have raised concerns about the dangers to human, terrestrial, and aquatic life. Goolsby et al. (1991) found that atrazine exceeded allowable drinking water contaminant levels (MDS) in 27% of samples from the smaller tributaries and the lower Mississippi and Missouri Rivers. While a full understanding of the possible dangers of atrazine to human health is not available, many environmentalists and health groups have raised a warning flag. Agricultural research groups have been busy researching the problems although a full understanding is not yet available.

Hypoxia is a condition where waters are depleted of their oxygen content and results in serious reduction of biological activity in the water. A zone approximately 18,000 km^2 in area, within the Gulf of Mexico near the mouth of the Mississippi River, has developed into a hypoxic zone with decreasing fish and shellfish densities. The zone is believed to be related to the concentrations of N, P, and silicate in the water. Since agriculture is believed to be the largest source of N and P in the river, agriculture has often been pinpointed as an industry that needs to use corrective measures.

Status and Trends in Soil and Water Conservation

Research and education have contributed greatly to development of sound public policy during the past century. Scientific and professional societies have contributed in exchange of information and in bringing research information to the attention of decision makers. But as conditions change, new problems arise. The population increase has intensified the need for better use and management of our natural resources. The multiple needs for land puts new pressures on the need for sound, quantitative research information. Increased research and education programs, including transfer to policy makers, in the future are imperative.

One could debate whether government policy shaped the nations research agenda, or that research led the way in guiding legislative policy development. A debate on these issues, however, would undoubtedly lead to the conclusion that they were developed hand-in-hand, as indeed they should. Certainly research has been indispensable in guiding government policy, and at times the demands for information by legislative and public groups have helped guide research agendas. In any event, now is an important time for researchers and policy makers to work closely together. It is an important time to inventory the state of our knowledge for conserving soil and water resources and to address the challenges facing the development and implementation of conservation practices on the land in the twenty-first century. This statement was the guiding principle for the authors invited to share their expertise in soil and water conservation as contributors to this book. Their thoughts and insights into various dimensions of soil and water conservation are presented in the following chapters.

References

Betts, E.M. *Thomas Jefferson's Farm Book* (with commentary and relevant extracts from other writings). American Philosophical Society and Princeton University Press. Princeton, NJ, 1953.

Colacicco, D., T. Osborn, and K. Alt. Economic Damage from Soil Erosion. *Journal Soil and Water Conservation* 44:35–39, 1989.

Goolsby, D.A., R.C. Cooper, and D.J. Markovchick. (1991). Distribution of Selected Herbicides and Nitrate in the Mississippi River and Its Major

Tributaries. April–June 1991: U.S. Geological Survey Water-Resources Investigations Report 91-4163, Denver, Colorado.

Holden, L.R., J.A. Graham, R.W. Whitmore, W.J. Alexander, R.W. Pratt, S.K. Liddle, and L.L. Piper. Results of the National Alachlor Well Water Survey. *Environmental Science and Technology* 26:935–943, 1992.

Pierce, F.J., R.H. Dowdy, W.E. Larson, and W.A.P. Graham. Soil Productivity in the Corn Belt: An Assessment of Erosion's Long-Term Effects. *Journal Soil Water Conservation* 39:131–138,1984.

Power, J.F. and J.S. Schepers. Nitrate Contamination of Groundwater in North America Agriculture. *Ecosystems and Environment* 26:165–188,1989.

Turner, E. and N. Rabalais. Changes in Mississippi River Water Quality This Century: Implications for Coastal Foods Webs. *BioScience* 41:140–147, 1991.

Understanding and Controlling Soil Erosion by Rainfall

John M. Laflen

Introduction

In 1813, Thomas Jefferson in a letter to C. W. Peale advocated "horizontal plowing": "Our country is hilly and we have been in the habit of ploughing in straight rows whether up and down hill, in oblique lines, or however they lead; and our soil was all rapidly running into the rivers. We now plough horizontally, following the curvatures of the hills and hollows, on the dead level, however crooked the lines may be. Every furrow then acts as a reservoir to receive and retain the waters, all of which go to the benefit of the growing plant, instead of running off into the streams. In a farm horizontally and deeply ploughed, scarcely an ounce of soil is carried off from it." Jefferson's 1813 letter demonstrates his awareness of the on- and off-site effects of soil erosion, the role of runoff in soil erosion, and the interaction of soil conservation, hydrology and crop production. These are still important topics today, nearly 200 years later.

Bennett (1939) in his treatise on soil conservation, covered nearly every item of current interest related to soil erosion, describing erosion processes, impacts of erosion in the U.S. on land degradation, off-site damages, and the impact of soil erosion on civilizations. Bennett's work was and is an undeniable force in our society. And, because of the leadership of the U.S. in soil and water conservation, Bennett's work has contributed to our ability to feed, clothe and house a world population far in excess of any that Bennett likely imagined.

Soil erosion research in the United States has been an evolving work. This chapter describes this scientific evolution in terms of major events since 1900 with regard to soil erosion by rainfall and its control. The focus will be on past events which provide the basis for present understanding and current research directions that establish the needed technology to conserve soil resources for future users.

Recognition of Erosion as a National and Global Problem

Little research in soil erosion would have been conducted if it had not been recognized as a serious problem. While early farmers such as Jefferson (1813) recognized erosion as a problem, it was not publicly recognized as a national problem until H. H. Bennett called it to the public's attention. H. H. Bennett, and his able assistant W. C. Lowdermilk, laid the groundwork for public support of soil erosion as a "menace to the national welfare." In 1930, Congress appropriated $16,000 for research into the causes of soil erosion and the preservation of soil and the prevention of erosion. Bennett (1939), in the preface to his book *Soil Conservation* attributed the educational campaign of the United States Department of Agriculture and the publication in 1928 of the United States Department of Agriculture Circular, "Soil Erosion—A National Menace," as critical elements in securing public and political attention to this menace.

While Bennett stirred the home fires, Lowdermilk surveyed ancient and modern land use in European and Mediterranean countries, with respect to erosion control, soil and water conservation, and flood control. Lowdermilk's (1940) regular written reports to Bennett are fascinating reading about victories and defeats in the war against soil erosion, while another war was beginning all around him. Lowdermilk's 1938-39 survey was summarized in a United States Department of Agriculture bulletin, "Conquest of the Land Through Seven Thousand Years." In that bulletin, Lowdermilk wondered if Moses might have delivered an eleventh commandment to establish man's relation to the earth, completing the trinity of responsibilities to his Creator, to his fellow man and to the holy earth (Lowdermilk, 1953).

Bennett and associates initiated public educational programs that eventually led to the recognition of erosion as a national problem. Others recognized the seriousness of soil erosion and the importance of soil conservation. McDonald (1941) described a century of early American soil conservationists ranging from the mid 1700s to the mid 1800s. Soil conservationists in this period discussed soil-building crops, liming, plowing on the contour, healing of gullies, wind erosion and even the soil erosion process. However, despite the recognition by farmers that soil erosion was a serious problem and despite the fact that European scientists had studied and published about soil erosion, the soil erosion problem had escaped the notice of U.S. research scientists (Nelson, 1958).

The Establishment of a Network of Erosion Research Stations

The period from 1930 to 1942 was the "golden years for conservation research" (Nelson, 1958). Erosion research stations were established representing ten major regions of the United States. These stations were located at Guthrie, OK; Temple, TX; Hays, KS; Tyler, TX; Bethany, MO; Statesville, NC; Pullman, WA; Clarinda, IA; LaCrosse, WI; and Zanesville, OH. Plot design was based on the studies by M. F. Miller and Associates at the University of Missouri (Meyer and Moldenhauer, 1985).

The formation of this network of erosion research stations, plus others initiated later, has provided a repository of research data that has been accessible and used by many scientists. These measurements provided a basis for the selection of practices from the 1930s to the 1950s. The present values of soil erodibility are anchored firmly to a set of "benchmark" soils, nine of which were established on the original ten sites above (Olson and Wischmeier, 1963; Wischmeier et al., 1971). The cropping-management factor in the Universal Soil Loss Equation (Wischmeier and Smith, 1961, 1965, 1978) was based almost entirely on measurements and interpretations from this immense data set, augmented by other similarly collected data (Wischmeier, 1960). The Pullman, WA site is the only one of the original network of ten that is functional as an erosion research station today.

Other closely related works took place because of these plots. One of the early rainfall simulation experiments to determine the effect of slope and length on soil erosion was performed by Duley and Hays (1932) and Duley and Ackerman (1934). Duley had worked with Miller at Missouri (Duley and Miller, 1923) and compared his experimental works with those of his former Missouri associates. Duley also published a classical study in 1939 on surface factors affecting the rate of intake of water by soils, concluding that "the intake of water is primarily controlled by the condition of the immediate surface, and the circumstances that may affect surficial changes," an observation still pertinent today.

Erosion plot data from the original and later plots are still used today for various purposes. The Water Erosion Prediction Project (WEPP) validation has drawn on historical data from two of the ten original sites—Guthrie, OK and Bethany, MO (Zhang et al., 1996), plus data from six other sites established with similar plot designs. Additionally, efforts are continuing to perform validation on a number of other sites, some of which were in the original ten. These data today are archived and are frequently accessed by scientists on the World Wide Web. They are available from the United States Department of Agriculture Agricultural Research Service, National Soil Erosion Research Laboratory, Purdue University, West Lafayette, Indiana.

The Establishment of a Scientific Basis for Soil Loss Limits

A linchpin of the soil conservation movement in this country, now codified into state and federal law, is the concept of an allowable (tolerable) soil loss limit which if not exceeded will not cause a productivity loss. This concept has great significance in the United States as it forms the scientific basis for many of our erosion programs, and because this concept led to the development and maintenance of a research program related to soil erosion prediction. Where did the concept of soil loss tolerance come from?

Bennett in 1939 distinguished between "normal erosion" and erosion caused by "human interference." "Normal erosion" was the "transposition of surface soil by rain, wind and gravitational movement at a pace no more rapid, generally, than the pace at which new soil is formed from the parent materials beneath;" however, Bennett quantified this concept no further.

Smith (1941) seems to have been the first to quantify a tolerable soil loss limit. In his paper he expresses a factor, A_l, as the soil loss with a given practice, and expressed it as $A_l = A\ P$, where A is predicted soil loss using an empirical equation and P is the effect of management practices. Smith describes A_l as a constant with a value of 4 tons per acre, equivalent to 250 years to erode 7 inches of surface soil. Smith further states that "the value of A_l should be based upon that rate of soil loss which will permit at least a constant or preferably an increasing time gradient of soil fertility." Smith discussed cases where A_l might not be equal to 4 tons per acre, but he did not generalize the concept to conservation planning on the larger scale, focusing only on the Shelby soil in Missouri.

Smith's 4 tons per acre as an allowable limit for the Shelby soil is quite similar to the tolerable soil loss limit for many of the nation's soils, and is certainly far greater (by at least a factor of 10) than the tolerable soil loss limit that would have been established based on Bennett's concept of soil erosion being equal to soil building from parent materials. The work in Africa by Owens and Watson (1979) showed these rates were in the order of fractions of tons per acre. Since Smith was a driving force in the Soil Conservation Service's research and a principal scientist involved in the development of the Universal Soil Loss Equation, it is highly likely that present-day tolerable soil loss limits began with this work.

Description of the Fundamental Processes in Erosion

A continuing focus of soil erosion research has been on the fundamental processes in soil erosion, which were well established when Nelson (1958) expressed his belief that fundamental research had been neglected for a series of reasons, and that more effort should have been put into fundamental research before the research stations were established.

Duley and Miller (1923), in interpreting experimental results and in extrapolating them to other soils, topography and climate, relied very little on an understanding of the fundamental processes of erosion, although they expressed an appreciation of the role of deposition at the foot of slopes in reducing sediment delivery. They had no opinion as to the effect of slope length on soil erosion per unit area, and had little grasp of the influence of slope. Yet by 1939, Bennett was describing sheet erosion, rill erosion and gully erosion.

The most prolific writer in terms of erosion processes in the last 100 years was W. D. Ellison (Ellison, 1944a-b,1945, 1947a-g, 1952; Ellison and Ellison,1947). His writings were first published in the mid 1940s when he was an agricultural engineer with the United States Department of Agriculture's Soil Conservation Service and last published in 1952, then as a consulting engineer with the Department of the Navy, Bureau of Yards and Docks. Ellison's statement that "soil erosion is a process of detachment and transportation of soil materials by erosive agents" (Ellison, 1947a) set the stage for much of our understanding and modeling of soil erosion processes. Meyer and Wischmeier (1969) explicitly developed a model based on these principles.

Ellison also described erosion studies designed to separate the effects of rainfall and flowing water on soil erosion (Ellison, 1945). He explicitly separated raindrop effects from runoff effects and discussed these separately. Water was added to initiate flow at the upper end of plots, with and without rainfall, to study the effects of rainfall and runoff separately and in combination. Such procedures are still in common use today (Laflen et al., 1991b).

The fundamental work and thinking in the early years has provided considerable guidance for the work of today. While Nelson was very concerned about the fact that little fundamental research was conducted before the erosion stations were developed, the work by Ellison and others was surely prompted by the results obtained from a wide range of experiments on different soils and topographies and in different climatic regions. Much of the fundamental work was surely stimulated to provide unifying concepts to rationalize disparate data sets. It is highly doubtful if the fundamental work would have been stimulated without these data sets. The question of which came first, the theory or the empiricism, has little merit in the continuing evolution of soil erosion research and control.

Development of Erosion Prediction Technology

Zingg (1940) presented a so-called "rational equation" in which total soil loss was expressed as a nonlinear function of land slope or horizontal length of land slope. He combined these into a single equation (Table 1). Zingg indicated that the exponents on slope and length and the coefficient C′ were functions of infiltration rate, physical properties of the soil, intensity and duration of the rain, and other factors.

Smith (1941) built upon Zingg's equation in formulating a more complete prediction equation (Table 1). Since he was working on one soil, and in one climatic region, his results had a limited area of application; however directions taken by Smith (1941) were clearly integrated into later prediction methods. For example, Smith defined the factor P as the ratio of soil loss with a mechanical practice to the soil loss without the practice—and these practices included contouring, rotation strip cropping and terracing. Additionally, Smith evaluated the C′ factor of Zingg for conditions at Bethany, MO, on the Shelby soil. Conditions evaluated included rotations, surface and subsoil soils, and the use of lime and fertilizer. Smith solved the rational equation, presenting figures for selection of cropping, management and conservation practices for a wide range of slopes and lengths.

Browning and coworkers (1947), building on the work of Zingg and Smith, developed a system of erosion prediction for Iowa conditions, and specifically

Table 1. Soil Erosion Prediction Equations

Zingg, 1940	$A = C'$		$L^{.0.6}$	$S^{1.4}$		
Smith, 1941 $A = C'''$			$L^{0.6}$	$S^{1.4}$		P
Browning, 1947	$A = C'''$	K'	$L^{0.6}$	$S^{1.4}$		P
Musgrave, 1947	$A' = P_{30}^{1.75}$	K''	$L^{0.35}$	$S^{1.35}$	C*	P
USLE, 1965	$A = EI_{30}$	K	$(L/72.6)^{0.5}$	$(.065 + .045\ S + .0065\ S^2)$	C	P
USLE, 1978	$A = EI_{30}$	K	$(L/72.6)^{0.5}$	$(65.4\ \mathrm{Sin}^2\Theta + 4.56\ \mathrm{Sin}\Theta\ \ +.065)$	C	P
MUSLE	$A = (Qq_p\ )^{.56}$	K	$(L/72.6)^{0.5}$	$(65.4\ \mathrm{Sin}^2\Theta + 4.56\ \mathrm{Sin}\Theta\ \ +.065)$	C	P
RUSLE, 1995	$A = EI_{30}$	K	$(L/72.6)^{m}$	$a\ \mathrm{Sin}\ \Theta + b$	C	P

A-Soil loss in tons/acre
A'-Soil loss in inches/year
C′,C″, C‴-Coefficients
C*-Vegetal cover factor
P_{30}-Maximum precipitation amount (inches) falling in 30 minutes in a storm
K′,K″,K-Soil erodibility factors
L-Slope length in feet
S-Slope in percent
Θ-Slope angle in degrees
C-Cropping management factor
Q-Storm runoff volume in inches
q_p-Peak runoff rate in inches per hour
E-Storm rainfall energy in hundreds of foot-tons per acre
I_{30}-Maximum rainfall intensity in a 30 minute period within a storm in inches per hour
P-Conservation practice factor
m-Exponent on length term-values depend on slope or slope and rill/interrill ratio
a,b-Coefficients in function making up slope term-values depend on slope

included a soil erodibility factor and specific permissible soil loss limits for a number of Iowa soils.

Musgrave (1947), writing on the results of work by a committee which included Smith and Browning, included an expression for the effect of rainfall on soil erosion in his prediction of soil loss (Table 1). Additionally, a first attempt to account for soil erodibility was made. Values for soil erodibility were established, adjusted to a 10%, 72-foot-long plot (a plot 6 feet wide and 72 feet long was .01 acre, simplifying computations before the days of electronic calculators and computers). These values of soil erodibility were different than but well correlated with those of the benchmark soils of Wischmeier and Smith (1978). In fact, the prediction technology advanced by Musgrave and coworkers was widely used until the Universal Soil Loss Equation- USLE (Wischmeier and Smith, 1961, 1965, and 1978) was fully implemented—well into the 1960s. The Musgrave equation as advanced in 1947 expressed soil loss in terms of inches, rather than the more common dimensions of tons per acre.

The USLE is a factor-based empirical equation derived from much erosion plot data. The USLE expresses soil loss as:

$$A = R\ K\ LS\ C\ P \qquad (1)$$

where A is predicted soil loss, R is a rainfall factor, K is a soil erodibility factor, LS is a topographic factor, C is a cropping and management factor and P is a conservation practice factor. The USLE is the current erosion prediction technology used over much of the world today.

As demonstrated in Table 1, the USLE was a natural evolution in erosion prediction technology. In fact, the Musgrave equation was similar in form to the USLE and many of the factor values were quite similar. The USLE development also had several workers, including D. D. Smith, that were involved in the development of the earlier technology. The USLE technology is widely used to assist in making decisions about management of lands to control sheet and rill erosion, but owes much of its form to the earlier work, and in fact, many of the same approaches taken in the development of the USLE were used earlier in the technologies of Smith, Browning and Musgrave.

However, while the USLE and earlier equations were apparently very similar, there were some very significant differences. A major difference involved basing the cropping and management practice factor used in the USLE on cropping periods, while those expressed up to that point were based on annual values. Annual values are used at the field level now in the USLE but they are developed using the cropping period C values, rather than on an annual "C" factor. This was carried one step further in a recent revision of the USLE (RUSLE, Renard et al., 1991) where C values are computed for 15-day periods. The significance of this difference was that the USLE factors for a period for several cropping systems could be used to develop new period C factors for different crops and management systems.

The USLE is a "Universal" equation because a rainfall factor—the product of a storm's rainfall energy and the maximum 30-minute rainfall intensity in the storm (Wischmeier and Smith, 1958) was developed which applied across most regions of the U.S. Up to that point, most erosion prediction procedures were somewhat regional in nature.

The USLE after release underwent several revisions (Wischmeier and Smith, 1965, 1978; Renard et al., 1991), to extend the USLE to more comprehensive sets of managements. The USLE became the erosion prediction technology of choice for much of the world because of the soundness of the factor values, because it was supported by the United States Department of Agriculture, and because it was user-friendly. Wischmeier, in a personal conversation with the author several years ago, alluded to the many interactions ignored (but known) in the USLE because it was necessary to have a user-friendly tool; to consider the many interactions would have required a tool too complex to use at the field level. RUSLE (Renard et al., 1991) considers many of these interactions and is a user-friendly tool when implemented on a computer.

Erosion prediction technology has advanced from the empirical USLE technology to a modeling technology usable at the field level called WEPP (Water Erosion Prediction Project), a product of a team of federal and state scientists (Laflen et al., 1991a). WEPP is a model that computes the state of the land, water and biomass system daily, and then takes a fundamental approach to computing runoff (the driving force in detachment by flowing water), and sediment detachment, transport, and deposition. WEPP is a erosion prediction system that employs the erosion processes described by Bennett and studied by Ellison, and links them with other processes that occur on the land to make a complete erosion prediction system. Erosion prediction technology has moved from the empirical to the fundamentals—but it was 55 years from the time Zingg published his equation to the release of WEPP for public use. The new technology can only be implemented because of the ubiquitous nature of the personal computer.

Development of Rainfall Simulation for Erosion Research

A major innovation in erosion research was the rainfall simulator, developed by L. D. Meyer (1965), which approximated the drop size distributions and fall velocities of natural rainfall. This innovation opened the field of soil erosion research to fundamental studies where rainfall could be controlled. Also, because of its dependability and low cost relative to erosion plots, rainfall simulation enabled soil erosion research by more scientists, and made it possible to study and evaluate cropping practices, land treatments, and erosion processes in less time. Meyer's critical contribution, generally accepted among erosion scientists, was the documentation of the characteristics of the rainfall from the nozzles that are used for rainfall simulators in much of the world (Meyer and Harmon, 1979; Foster et al., 1979; Swanson, 1965; Norton and Brown, 1992).

Meyer was not the first to perform rainfall simulation. The slope factor in Zingg's original equation (Table 1) was based on two rainfall simulation studies. One was by Duley and Hays (1932) using a "sprinkling can method," exactly as one would picture it, people walking holding a can that was sprinkling on the soil surface (for a picture of this work, see Duley and Ackerman, 1934).

Rainfall simulators were important in quantifying factors in the USLE and studying erosion processes. The data used to develop the soil erodibility nomograph (Wischmeier et al., 1971) was collected using a rainfall simulator; the effect of conservation tillage on soil erosion in the USLE and RUSLE came from rainfall simulation studies (Wischmeier, 1973; Swanson et al., 1965; Siemens and Oschwald, 1978; Laflen et al., 1978). The effect of crop residue on flow velocity was studied by Kramer and Meyer (1969); the role of raindrop impact in soil detachment and transport was studied by Young and Wiersma (1973); and more recently, interrill erosion processes were analyzed (Meyer and Harmon, 1992). Any serious erosion research scientist today has at least one, and perhaps more, rainfall simulators available for use in their studies.

Development of Erosion Models

Erosion modeling has been of interest for some time. The first erosion model that explicitly incorporated the concepts of Ellison was the model of Meyer and Wischmeier (1969). This model incorporated both rill and interrill detachment and transport, and used a transport or detachment limiting concept. This model did not explicitly deal with deposition, but if the transport capacity was less than the material being transported, the excess material was assumed to be deposited. The model was a steady state model, operating at a single flow rate (or excess rainfall rate) and operated on a single storm basis.

One of the most unique models is the EPIC model (Williams et al., 1984; Williams, 1993). The model was developed explicitly to predict the effect of soil erosion on crop productivity, and includes components related to hydrology, crop growth, soil erosion and sediment removal. The erosion component is not sophisticated. It includes several empirical approaches to erosion and sediment yield. These components are either the USLE or a closely related technology. EPIC was used in making estimates of the effect of erosion on the nation's land's ability to produce food and fiber. The technique was to predict erosion, to simulate the effect of soil removal on the properties of the soil and its profile, and to predict crop yield from this changed profile.

The definitive work on modeling the erosion process is that of Foster (1982). Foster led the development of the erosion and sediment yield portion of CREAMS (United States Department of Agriculture, 1980, Foster et al.,

1981), with much of the modeling approach described in his work of 1982. CREAMS linked hydrology, erosion, and sediment, nutrient, and pesticide delivery from small watersheds.

In 1985, the United States Department of Agriculture initiated a project to replace the USLE for predicting soil erosion (Laflen et al., 1991a) that led to the WEPP model. From the development of the first model of Meyer and Wischmeier (1969) to full implementation of WEPP (Water Erosion Prediction Project) will require something in the order of 30 years, a time frame very comparable to that from the development of the first erosion prediction technology in 1940 (Zingg) to full implementation of the USLE in the mid to late 1960s. Important milestones have been the development of a set of user requirements (Foster, 1987) for WEPP, release of the model in 1989, and the release of the model for public use in 1995, a decade after a workshop recommended such a venture be initiated.

Control of Soil Erosion

The reason for trying to understand erosion processes and for predicting soil erosion is to efficiently and effectively control soil erosion. Jefferson's interest in contouring was an early example of the recognition of the problem and an attempt at solutions. Ancient civilizations had practiced erosion control. In the United States, early soil conservationists recognized erosion as a problem and developed techniques that enriched the soil, increased yields and reduced soil erosion (McDonald, 1941). These practices included liming, drainage, cover crops, sod-based rotations, and contouring.

The major advances in erosion control that have occurred in the last 100 years are few in number but great in the area of application. The principal advances that are now being applied are conservation tillage systems and chemical control methods. Additionally, materials have been developed specifically for control of soil erosion, and enhancements in effectiveness have occurred in almost every erosion control method.

The principles in erosion control have been well developed over the last century, and are used in selecting effective erosion control practices (Meyer et al., 1991). Erosion by rainfall is caused when the forces applied by rainfall and runoff water exceed the resisting forces and detachment occurs. If energy for sediment transport is available, the detached sediment can be transported from the site and erosion has occurred. Erosion control efforts have focused on reducing the forces available to detach soil, increasing the resistance of soil to detachment, and reducing the energies available for sediment transport. These are easy principles to grasp, but difficult to implement on land surfaces where the management is very dynamic, where funds to implement control are limited, and where the forces and resistances are very dynamic and subject to wide variations in time and space.

Early control methods focused, and many current methods focus, on controlling soil erosion by managements that reduce runoff and soil erodibility. Controls would include the recent use of grass on highly erodible land and the earlier use of sod-based rotations. Crop rows on the contour can reduce both the forces of detachment by flowing water and the energies to transport sediment. Diversions and terraces can perform the same function. Terraces with underground outlets (Ramser, 1917) can effectively discharge runoff water from a field in a channel (a tile) that is nonerodible. Grass waterways reduce flow velocity, inducing deposition. Additionally, the resistance to detachment in a grass waterway is sufficient to eliminate detachment within the waterway for all but rare storms.

The real problem in erosion control is not having an array of practices that can control soil erosion, but having practices that meet the need for economic and soil sustainability. Recently, two techniques—conservation tillage and chemical control—have helped to meet these two needs.

Some form of conservation tillage (defined here as having more than 30% of the land surface covered by residue after planting) is used on 35% of all planted acres in the United States (CTIC, 1994). The adoption of conservation tillage for crop production has overcome the criticism of conservation programs advanced by Schultz (quoted by Wolman, 1985) that "programs are unsuccessful primarily because no profitable rewarding new agricultural inputs have been available to farmers which they could adopt and use." As Allmaras et al. (1985) noted, "American agriculture can indeed gain profitability and greater conservation from conservation tillage." A recent series of regional United States Department of Agriculture-ARS reports (Black and Moldenhauer, 1994; Blevins and Moldenhauer, 1995; Langdale and Moldenhauer, 1995; Moldenhauer and Mielke, 1995; Papendick and Moldenhauer, 1995; and Stewart and Moldenhauer, 1994) have highlighted crop residue management to reduce erosion and improve soil quality for most regions of the United States.

Conservation tillage has truly been a revolution in tillage on the American farm. While tillage defined as "conservation tillage" covers more than 35 % of U.S. cropland, an additional 26% of cropland was not plowed in 1994. In the 1960s, there was virtually no cropland in the U.S. that was unplowed, and tillage equipment and herbicides for such conditions were barely available. Even as late as 1985, Allmaras et al. stressed the need for development of effective machinery and herbicides. Now, these pieces have been developed, and the moldboard plow, the former tillage tool of choice for much of agriculture, has virtually disappeared in many places, all in less than a quarter of a century. In many places, conservation tillage has replaced sod-based rotations, terracing and contouring as the technology of choice for erosion control. Under some conditions, conservation tillage does not sufficiently protect the soil and additional practices are required to reduce soil erosion and sediment delivery.

To a much lesser degree, but potentially very promising, is the increasing use of chemicals to stabilize the soil surface for erosion control. Polyacrylamide (PAM) has been studied to ameliorate soil conditions and improve yields (Wallace and Wallace, 1986 a,b) and increase water infiltration (Terry and Nelson, 1986; Mitchell, 1986). Wallace and Wallace (1986c) also showed that PAM and fly ash could be used in combination when applying fly ash to the soil so as to maintain an acceptable soil physical condition. Wallace and Wallace (1986d) showed that PAM could reduce soil erosion greatly, but rates used in their experimental work ranged from about 5 up to 134 kg ha^{-1}. They suggested that PAM might be applied in very low amounts (as low as a kg ha^{-1}), although most of their work was at much higher application levels. Shainberg et al. (1994) discovered that a thin layer of polyacrylamide (PAM) at the soil surface could increase the stability of the soil surface, reducing its failure due to flowing water. Shainberg, working with United States Department of Agriculture scientists at Kimberly, ID (Lentz et al., 1992), tested PAM by adding it at very low concentrations (less than .002 %) and low amounts (1.11 kg ha^{-1} or less) in furrow irrigation and found that it reduced soil erosion by 22% to 80% in the first four irrigations after application (no cultivation after application). Trout et al. (1995) found that PAM at low concentrations "dramatically reduces furrow erosion and sediment loss." PAM is commercially available and is being used in irrigated agriculture in the Western U.S. (Trout et al., 1995).

Flanagan et al. (1992) studied PAM and power plant gypsiferous by-products under simulated rainfall and runoff on a silt loam soil in Indiana. They concluded that these additives could be used to reduce soil erosion and increase infiltration under rainfall conditions on many soils in the U.S. While PAM rates of 20 kg ha^{-1} were much higher than that used by Trout et al. (1995), the potential under rainfall conditions was amply demonstrated.

Frontiers in Soil Erosion Research

What are the new frontiers to be explored? Much sharpening of existing practices must still continue, basic knowledge into the erosion processes must still be gained, and the development of decision-making technologies-modeling will continue. But what areas of research will produce the new science needed to ensure a sustainable soil resource to meet the needs of mankind? Many are terribly concerned that we'll not be able to feed the world in the next century. In 2098, what will be written about the great research findings that have helped sustain the soil resource and benefited mankind?

Three major advances will highlight such writing in 2098: (1) the development of strategies and practices for the control of surface runoff, (2) the development of economically viable production systems that control erosion, and (3) engineering of plants for production and protection.

Strategies and practices that reduce runoff are required to control the land degradation that accompanies severe erosion. Gullying destroys land, is caused by high volumes and rates of runoff, and is a worldwide problem. Science must provide answers and technology must be developed to reduce surface runoff. These technologies will include those that keep the soil porous and allow high infiltration rates, and promote water movement to at least the maximum rooting depth of most crops. Such practices will increase water availability to crops, reduce drought frequency and severity, improve stream flow, recharge groundwater, and reduce rill, channel and gully erosion. Appropriate technologies may include chemical treatments such as PAM for control of the soil surface, but could include biological remediation of soil profiles such as might be achieved by existing or new plants engineered to flourish in soils having unfavorable root zones. In 2098, science will have further developed the use and management of living organisms for bioremediation of the soil profile. In some areas, physical remediation, such as tested by Bradford and Blanchar (1980) on a Missouri fragiudalf, might be an acceptable means of restoring production and reducing runoff. In 2098, technologies that reduce surface runoff will have made a tremendous difference in many regions of the U.S. and in much of the world.

Economically viable production systems that control erosion are imperative if we are to control erosion and the accompanying land degradation on a broad scale. Engineering of plants to reduce water use may make it practical to produce crops on land where the erosion hazard due to high rainfall is low, reducing the demand for land that is most susceptible to soil erosion. The reduction of soil erosion on land in the conservation reserve program (CRP) has been significant. The development of an economically viable cropping system for such lands might make it possible to use them over the long term in a resource-conserving manner, avoiding the need for government support for nonerodible uses of these lands. Conservation tillage exemplifies such a system. Various forms of conservation tillage have been developed for much of the world and their development to meet local and regional situations must be continued. In 2098, improvements in production systems will raise the standard of living of farmers, provide for a more stable food supply, and reduce soil erosion and land degradation in much of the world.

Plants can be engineered to provide both production and protection. Plants provide considerable protection of the soil surface, using soil water (which reduces runoff) and intercepting raindrop splash. Plant residues reduce runoff velocity and provide protection from hydraulic shear and raindrop splash. Plants provide above- and belowground biomass which can maintain or improve the organic matter content of the soil, thereby increasing soil water-holding capacity, increasing soil strength (Barry et al., 1991), and increasing nutrient availability. Fundamental research into the nature of these organic materials and their effect on soil strength must continue, along with engineering of plants to provide the best organic materials and to provide for their long-lived protection of the soil resource. In the year 2098, engineered plants will provide much-needed plant protection to conserve the soil resource.

Many areas of research are equally important. Fundamental research is needed to describe the forces that detach and those that resist detachment of soil. Modeling of landscape processes will help us identify where some treatments are effective and where others are not, and will also help us identify where our research can have an impact, and where it may not.

While my focus is on soil erosion research, the fact is that soil erosion control practices must be implemented to control erosion. Conservation tillage could not be implemented without the proper machinery and chemicals. Research and development in enabling technologies and their agronomic feasibility will be required in order to implement new erosion control and management practices that may sustain the nation's lands.

Conclusion

The production of food and fiber is a basic requirement of a modern civilization. High quality, carefully managed, soil, land and water resources are essential to sustain the human population envisioned for the planet earth, for what we do to our soil impacts not only the production of food and fiber, but the quality of the life that the soil sustains. Degradation of the soil resource must not only be halted but reversed, a formidable task requiring major research and technology development efforts.

The U.S. has a fantastic heritage in soil erosion and conservation research and a good record of implementation of the developed technologies on the world's lands. Our heritage suggests that land use and management are very dynamic, that new production technologies will emerge, that new crops will be developed, that new needs to control erosion will evolve and that new solutions for erosion control must and will be found. Our heritage from the last 100 or so years is that of good public support for a healthy, dynamic and motivated scientific community responsive to the need for a well-managed land base that will continue to meet the needs of mankind. To meet the needs of the next century, this heritage must continue.

References

Allmaras, R. R., P. W. Unger, and D. W. Wilkins. 1985. Conservation tillage systems and soil productivity, pp. 357–412. In: R. F. Follett and B. A. Stewart, eds. Soil Erosion and Crop Productivity, American Society Agronomy, Madison, WI.

Barry, P. V., D. E. Stott, R. F. Turco, and J. M. Bradford. 1991. Organic polymers' effect on soil shear strength and detachment by single raindrops. Soil Sci. Soc. Am. J. 55(3):799–804.

Bennett, H. H. 1939. Soil Conservation. McGraw-Hill Book Company, Inc., New York, NY.

Black, A. L. and W. C. Moldenhauer, eds. 1994. Crop residue management to reduce erosion and improve soil quality: Northern Great Plains. United

States Department of Agriculture, Agricultural Research Service Conservation Research Report No. 38.

Blevins, R. L. and W. C. Moldenhauer, eds. 1995. Crop residue management to reduce erosion and improve soil quality: Appalachia and Northeast. United States Department of Agriculture, Agricultural Research Service Conservation Research Report No. 41.

Bradford, J. M. and R. W. Blanchar. 1980. The effect of profile modification of a fragiudalf on water extraction and growth by grain sorghum. Soil Sci.Soc. Am. J. 44(2):374–378.

Browning, G. M., C. L. Parish and J. Glass. 1947. A method for determining the use and limitations of rotation and conservation practices in the control of soil erosion in Iowa. J. Am. Soc. Agron. 39(1):65–73.

CTIC. 1994. Annual Report, Conservation Technology Information Center. W. Lafayette, IN.

Duley, F. L. 1939. Surface factors affecting rate of intake of water. Soil Sci. Soc. Am. Proc. 4(1):60–64.

Duley, F. L. and O. E. Hays. 1932. The effect of the degree of slope on runoff and soil erosion. J. Agric. Res. 45(6):349–360.

Duley, F. L. and F. G. Ackerman. 1934. Runoff and erosion from plots of different lengths. J. Agric. Res. 48(6):505–510.

Duley, F. L. and M. F. Miller. 1923. Erosion and surface runoff under different soil conditions. Missouri Agricultural Experiment Station Research Bulletin 63.

Ellison, W. D. 1944a. Studies of raindrop erosion. Agric. Eng. 25(4):131–136.

Ellison, W. D. 1944b. Studies of raindrop erosion. Agric. Eng. 25(5):181–182.

Ellison, W. D. 1945. Some effects of raindrops and surface-flow on soil erosion and infiltration. Trans. Am. Geophysical Union 26(3):415–429.

Ellison, W. D. 1947a. Soil Erosion. Soil Sci. Soc. Am. Proc. 12(5):479–484.

Ellison, W. D. 1947b. Soil erosion studies-part 1. Agric. Eng. 28(4):145–146.

Ellison, W. D. 1947c. Soil erosion studies-part II, soil detachment hazard by raindrop splash. Agric. Eng. 28(5):197–201.

Ellison, W. D. 1947d. Soil erosion studies-part III, some effects of soil erosion on infiltration and surface runoff. Agric. Eng. 28(6):245–248.

Ellison, W. D. 1947e. Soil erosion studies-Part IV, soil erosion, soil loss, and some effects of soil erosion. Agric.Eng. 28(7):297–300.

Ellison, W. D. 1947f. Soil erosion studies-Part V, soil transportation in the splash process. Agric. Eng. 28(8):349–351, 353.

Ellison, W. D. 1947g. Soil erosion studies-Part VI, soil detachment by surface flow. Agric. Eng. 28(9):402–405, 408.

Ellison, W. D. 1952. Raindrop energy and soil erosion. Empire J. Exp. Agric. 20(78):81–97.

Ellison, W. D. and O. T. Ellison. 1947. Soil erosion studies-Part VI, soil transportation by surface flow. Agric. Eng. 28(10):442–444, 450.

Flanagan, D. C., L. D. Norton and I. Shainberg. 1992. Water chemistry effects on infiltration, runoff and erosion. Paper No.92-2053. American Society Agricultural Engineers, St. Joseph, MI.

Foster, G. R. 1982. Modeling the erosion process, pp. 297–388. In: C.T. Haan, H.P. Johnson and D. L. Brakensiek, eds., Hydrologic Modeling of Small Watersheds, American Society of Agricultural Engineers, St. Joseph, MI. Monograph No. 5.

Foster, G. R. 1987. User Requirements: USDA-Water Erosion Prediction Project(WEPP). NSERL Report No. 1, National Soil Erosion Research Laboratory, United States Department of Agriculture-Agricultural Research Service, West Lafayette, IN.

Foster, G. R., F. P. Eppert, and L. D. Meyer. 1979. A programmable rainfall simulator for field plots. Proceedings of United States Department of Agriculture-Agricultural Research Service workshop on rainfall simulators. United States Department of Agriculture-Science and Education Administration. Report ARM-W-10, pp. 45–59.

Foster, G. R., L. J. Lane, J. D. Nowlin, J. M. Laflen, and R. A. Young. 1981. Estimating erosion and sediment yield on field-sized areas. Trans. Am. Soc. Agric. Eng. 24(5):1253–1262.

Jefferson, T. 1813. Letter to C. W. Peale, Apr. 17, 1813. p. 509. In: Thomas Jefferson's Garden Book, annotated by E. M. Betts. American Philosophical Society, Philadelphia, PA (1944).

Kramer, L. A. and L. D. Meyer. 1969. Small amounts of surface mulch reduce soil erosion and runoff velocity. Trans. Am. Soc. Agric. Eng. 12(5): 639–641, 645.

Laflen, J. M., J. L. Baker, R. O. Hartwig, W. F. Buchele, and H. P. Johnson. 1978. Soil and water loss from conservation tillage systems. Trans. Am. Soc. Agric. Eng. 21(5):881–885.

Laflen, J. M., L. J. Lane, and G. R. Foster. 1991a. WEPP-A new generation of erosion prediction technology. J. Soil Water Conserv. 46(1):35–38.

Laflen, J. M., W. J. Elliot, J. R. Simanton, C. S. Holzhey, and K. D. Kohl. 1991b. WEPP soil erodibility experiments for rangeland and cropland soils. J. Soil Water Conserv. 46(1):39–44.

Langdale, G. W. and W. C. Moldenhauer, eds. 1995. Crop residue management to reduce erosion and improve soil quality: Southeast. United States Department of Agriculture, Agricultural Research Service Conservation Research Report No. 39.

Lentz, R. D., I. Shainberg, R. E. Sojka and D. L. Carter. 1992. Preventing irrigation furrow erosion with small applications of polymers. Soil Sci. Soc. Am. J. 56(6):1926–1932.

Lowdermilk, W. C. 1940. Letters to H.H. Bennett. Available in Manuscript Collection 279, Special Collections Department, Iowa State University Library, Ames IA.

Lowdermilk, W. C. 1953. Conquest of the land through seven thousand years. Agricultural Information Bulletin No. 99. Washington, D.C.

McDonald, A. 1941. Early American soil conservationists. United States Department of Agriculture Miscellaneous Publication 449. Washington, D.C.

Meyer, L. D. 1965. Simulation of rainfall for soil erosion research. Trans. Am. Soc. Agric. Eng. 8(1):63–65.

Meyer, L. D. and W. C. Harmon. 1979. Multiple-intensity rainfall simulator for erosion research on row sideslopes. Trans. Am. Soc. Agric. Eng. 23(1):100–103.

Meyer, L. D. and W. C. Harmon. 1992. Interrill runoff and erosion: Effects of row-sideslope shape, rain energy and rain intensity. Trans. Am. Soc. Agric. Eng. 35(4):1199–1203.

Meyer, L. D. and W. C. Moldenhauer. 1985. Soil erosion by water: The research experience, pp. 192–204. In: Agricultural History. University of California Press, Berkley, CA.

Meyer, L. D. and W. H. Wischmeier. 1969. Mathematical simulation of the process of soil erosion by water. Trans. Am. Soc. Agric. Eng. 12(6): 754–758, 762.

Meyer, L. D., C. Wu, and E. H. Grissinger. 1991. Use of basic erosion principles to identify effective erosion control practices, pp. 124–132. In: W. C. Moldenhauer et al., (eds) Development of Conservation Farming on Hillslopes. Soil and Water Conservation Society, Ankeny, IA.

Mitchell, A. R. 1986. Polyacrylamide application in irrigation water to increase infiltration. Soil Sci. 141(5):353–358.

Moldenhauer, W. C. and L.N. Mielke, eds. 1995. Crop residue management to reduce erosion and improve soil quality: North Central. United States Department of Agriculture, Agricultural Research Service Conservation Research Report No. 42.

Musgrave, G. W. 1947. The quantitative evaluation of factors in water erosion-a first approximation. J. Soil Water Conserv. 2(3):133–138, 170.

Nelson, L. B. 1958. Building sounder conservation and water management research programs for the future. Soil Sci. Soc. Am. Proc. 22(4):355–358.

Norton, L. D. and L. C. Brown. 1992. Time effect on water erosion for ridge tillage. Trans. Am. Soc. Agric. Eng. 35(2):473–478.

Olson, T.C. and W. H. Wischmeier. 1963. Soil-erodibility evaluations for soils on the runoff and erosion stations. Soil Sci. Soc. Am. Proc. 27(5) 590–592.

Owens, L. B. and J. P. Watson. 1979. Rates of weathering and soil formation on granite in Rhodesia. Soil Sci. Soc. Am. J. 43(1):160–166.

Papendick, R. I. and W. C. Moldenhauer, eds. 1995. Crop residue management to reduce erosion and improve soil quality: Northwest. United States Department of Agriculture, Agricultural Research Service Conservation Research Report No. 40.

Ramser, C. E. 1917. Prevention of the erosion of farm lands by terracing. Agricultural Bulletin 512. United States Department of Agriculture, Washington, D. C.

Renard, K. G., G. R. Foster, G. A. Weesies and J. P. Porter. 1991. RUSLE: Revised universal soil loss equation. J. Soil Water Conserv. 46(1):30–33.

Shainberg, I., J. M. Laflen, J. M. Bradford and L. D. Norton. 1994. Hydraulic flow and water quality characteristics in rill erosion. Soil Sci. Soc. Am. J. 58(4):1007–1012.

Siemens, J. C. and W. R. Oschwald. 1978. Corn-soybean tillage systems, erosion control, effects on crop production, costs. Trans. Am. Soc. Agric. Eng. 21(2):293–302.

Smith, D. D. 1941. Interpretation of soil conservation data for field use. Agric. Eng. 22(5):173–175.

Stewart, B. A. and W. C. Moldenhauer, eds. 1994. Crop residue management to reduce erosion and improve soil quality: Southern Great Plains. United States Department of Agriculture, Agricultural Research Service Conservation Research Report No. 37.

Swanson, N. P. 1965. Rotating boom rainfall simulator. Trans. Am. Soc. Agric. Eng. 8(1):71–72.

Swanson, N. P., A. R. Dedrick, H. E.Weakly, and H. R. Haise. 1965. Evaluation of mulches for water-erosion control. Trans. Am. Soc. Agric. Eng. 8(3):438–440.

Terry, R. E. and S. D. Nelson. 1986. Effects of polyacrylamide and irrigation method on soil physical properties. Soil Sci. 141(5):317–320.

Trout, T. J., R. E. Sojka, and R. D. Lentz. 1995, Polyacrylamide effect on furrow erosion and infiltration. Trans. Am. Soc. Agric. Eng. 38(3):761–765, 766.

United States Department of Agriculture. 1928. Soil Erosion-A National Menace. United States Department of Agriculture Circular 33. Washington, D.C.

U.S. Department of Agriculture. 1980. CREAMS-A field scale model for chemicals, runoff, and erosion from agricultural management systems. Conservation Research Report 26, United States Department of Agriculture, Washington, D.C.

Wallace, A. and G. A. Wallace. 1986a. Effects of very low rates of synthetic soil conditioners on soils. Soil Sci.141(5):324–327.

Wallace, A. and G. A. Wallace. 1986b. Effect of polymeric soil conditioners on emergence of tomato seedlings. Soil Sci. 141(5):334–342.

Wallace, A. and G. A. Wallace. 1986c. Control of soil erosion by polymeric soil conditioners. Soil Sci.141(5):363–367.

Wallace, A. and G. A. Wallace. 1986d. Enhancement of the effect of coal fly ash by a polyacrylamide soil conditioner on growth of wheat. Soil Sci. 141(5):387–389.

Williams, J. R. ed., 1993. The Erosion-Productivity Impact Calculator, Volume I. Model Documentation, Agricultural Research Service, United States Department of Agriculture.

Williams, J. R., C. A. Jones, and P. T. Dyke. 1984. A modeling approach to determining the relationship between erosion and soil productivity. Trans. Am. Soc. Agric. Eng. 27(1):129–144.

Wischmeier, W. H. 1960. Cropping-Management factor evaluations for a Universal Soil-Loss Equation. Soil Sci. Soc. Am. Proc. 24(4):322–326.
Wischmeier, W. H. 1973. Conservation tillage to control water erosion. Proceedings National Conservation Tillage Conference. Soil and Water Conservation Society, Ankeny, IA.
Wischmeier, W. H., C. B. Johnson and B. V. Cross. 1971. Soil erodibility nomograph for farmland and construction sites. J. Soil Water Conserv. 26:189–193.
Wischmeier, W. H. and D. D. Smith. 1958. Rainfall energy and its relationship to soil loss. Trans. Am. Geophysical Union 39:285–291.
Wischmeier, W. H. and D. D. Smith. 1961. A universal equation for predicting rainfall-erosion losses-An aid to conservation farming in humid regions. United States Department of Agriculture, Agricultural Research Service. ARS Special Report 22-66.
Wischmeier, W. H. and D. D. Smith. 1965. Predicting rainfall-erosion losses from cropland east of the Rocky Mountains-Guide for selection of practices for soil and water conservation. United States Department of Agriculture, Agricultural Handbook No. 282.
Wischmeier, W. H. and D. D. Smith. 1978. Predicting rainfall-erosion losses-A guide to conservation farming. United States Department of Agriculture, Agricultural Handbook No. 537.
Wolman, M. G. 1985. Soil erosion and crop productivity: A worldwide perspective, pp. 9–21. In: R. F. Follett and B. A. Stewart, eds. Soil Erosion and Crop Productivity, American Society Agronomy, Madison, WI.
Young, R. A. and J. L. Wiersma. 1973. The role of raindrop impact in soil detachment and transport. Water Resour. Res. 9(6):1629–1636.
Zhang, J. C., M. A. Nearing, L. M. Rissee, and K. C. McGregor. 1996. Evaluation of WEPP runoff and soil loss prediction using natural runoff plot data. Trans. Am. Soc. Agric. Eng. 39(3):855–863.
Zingg, A. W. 1940. Degree and length of land slope as it affects soil loss in runoff. Agric. Eng. 21(2):59–64.

Understanding and Managing Irrigation-Induced Erosion

R.E. Sojka

Introduction

Preventing irrigation-induced erosion takes on a special importance because of the indispensable role irrigated agriculture plays in feeding and clothing humanity. Irrigation is one of humanity's most potent weapons in the war against starvation and one of the best strategies for preserving earth's remaining undisturbed environments while meeting human food and fiber needs.

Overall, irrigation is used on about one-sixth of both the U.S. (Figure 1) and global cropped area, but irrigated cropland produces about one-third the annual harvest in both cases (Table 1), and about one-half the value of all crops (food, fiber, etc.) harvested (Bucks et al., 1990). A mere 50 million irrigated hectares (125 million acres), or about 4% of the world's total cropped land, produces about one-third of the world's ***food harvest*** (Tribe, 1994). Irri-

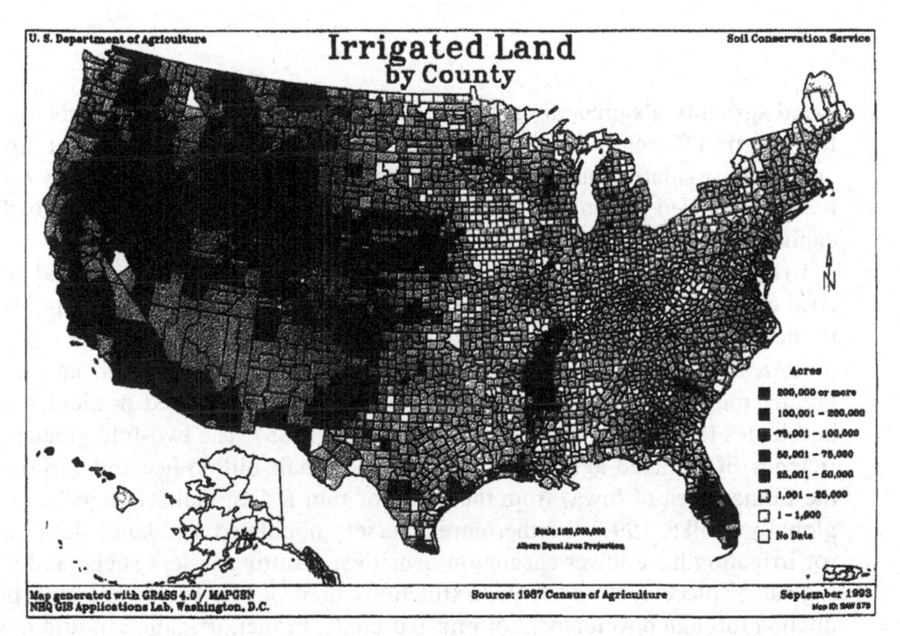

Figure 1. Survey of irrigated land by county in the U.S. (From the Soil Conservation Service, 1993).

Table 1. A General Summary of the Extent and Importance of Irrigated Agriculture Compiled from Several Sources[a]

Extent of Irrigated Crop Areas (ha)	
Worldwide Cropped Area (ha)	1.2–1.4 billion
Irrigated Crop Area	219–243 million (15-18 %)
U.S. Irr. Crop Area	24.1 million (14.8 %)
Surface Flow	13.2 million
Sprinkler	9.9 million
Other	1.0 million
17 Western State Total	18.8 million

Importance of Irrigated Agriculture

Irrigation occurs on one-sixth of global and US cropped area:

- producing 1/3 the annual global and US overall harvest
- producing 1/2 the monetary value of crops harvested
- producing 1/3 the global *food harvest* on the best 50 million ha
- freeing 1/2 billion ha of rainfed land in natural habitat
- providing greatly enhanced food security (yield reliability)
- providing (where surface waters are developed) flood control, transport, recreation, hydropower, community development

[a]Anonymous, 1995; Tribe, 1994; Hoffman et al., 1990; Gleick, 1993; Hunst and Powers, 1993.

gated agriculture's predominant association with aridity combines higher photosynthetic efficiency (fewer cloudy days) with the ability to prevent stress and better regulate inputs. These advantages result in higher commodity quality than rain-fed agriculture with greater yield assurance, often for crops that cannot be commercially grown otherwise.

Irrigated agriculture also provides substantial environmental dividends. Arid climates generally have lower disease and insect pressures, reducing pesticide requirements. Arid soils have high base saturation and little organic matter, thus they seldom require lime or potassium fertilizers, and have greatly reduced application rate requirements for soil-applied pesticides and herbicides (Stevenson, 1972; Ross and Lembi, 1985). The two-fold greater efficiency of irrigated agriculture frees almost a half billion hectares (36 times the farmed area of Iowa) from the need for rain-fed agricultural development globally (Sojka, 1997). Furthermore, sparsely populated arid lands developed for irrigation have lower speciation densities, resulting in less social and biological displacement and fewer extinctions than occurs for comparable production through development of rain-fed lands. In fact, irrigation of arid lands provides habitat that may extend the range of some humid and subhumid wildlife and avian species.

Table 2. Percent Maximum Yield of Portneuf Soil Having the Entire A Horizon Removed[a]

Crop	% Max. Yield Without A Horizon
Wheat	51
Sweet corn	52
Alfalfa	67
Dry Bean	60
Barley	68
Sugarbeet	79

[a]From Carter, 1993.

The Severity of Irrigation-Induced Erosion

Irrigation-induced erosion is a threat both to the sustainability of irrigated agriculture and to global food security. Yet, relatively little data have been published quantifying the extent of irrigation-induced erosion. Most published data originate from the U.S. Pacific Northwest (PNW), and focus largely on furrow irrigation. While not entirely representative of all irrigation-induced erosion, the data from this region demonstrate considerations common to nearly all irrigated agriculture.

Arid zone soils are usually low in organic matter and poorly aggregated, with thin, easily eroded A horizons. Carter (1993) demonstrated that, once eroded, yield potentials of PNW soils are severely reduced (Table 2). Furthermore, furrow irrigation, used on much of the world's irrigated land, is an inherently erosive process (Figure 2).

Furrow irrigation-induced erosion in the PNW commonly removes 5–50 tons ha^{-1} of soil per year, with much of the erosion (3–8x the field averaged rate) occurring near the upper end of fields near furrow inlets (Berg and Carter, 1980; Kemper et al., 1985; Fornstrom and Borelli, 1984, Trout, 1996). Over 50 t ha^{-1} soil loss has been measured for a single 24 hr irrigation (Mech, 1959). The magnitude of this problem is better appreciated when one recognizes that typical soil loss tolerance values for these soils are around 11 t ha^{-1} (5 tons/acre) per year. Thus, in the 90–100 years that PNW furrow irrigation has been practiced, many fields have little or no topsoil remaining on the upper one-third of the field. Furthermore, the topsoil remaining on lower field portions is mixed with subsoil washed off upper field reaches and deposited at the lower end.

The negative impacts of soil loss are numerous (Carter, 1990). The B horizons of most arid zone soils have poor chemical and physical properties. They easily crust, seal, and compact, and often have phosphorous and micronutrient deficiencies, which collectively impair emergence, fertility, rooting, absorption of water and nutrients, and yields. As yield potential decreases, input

Figure 2. Sediment deposition in a tailwater ditch from a single 12 hour furrow irrigation on Portneuf Silt Loam Soil on a 275 meter long 1.5% slope in Kimberly, Idaho. Erosional losses at upper furrow ends are typically 3–8 times the measured runoff sediment losses.

costs increase, while the probability of response from inputs declines. Thus, production cost increases while yield and profit decline.

Eroded soil deposits in the lower reaches of fields, and in drains, return-flow ditches, lakes, streams, or rivers. Even when a significant amount of sediment is captured in the lower reaches of the field or in containment ponds, redistribution onto the field is required. The societal costs of these losses include reduced net on-farm returns and reduced production, with resultant upward pressure on commodity pricing; higher cost of canal maintenance, river dredging, and algal control; riparian habitat degradation and biodiversity reduction; water contamination; impairment of fisheries and recreational resources; reservoir capacity reduction; and accelerated hydroelectric generator wear (Sojka and Lentz, 1995). Many of these expenses and losses are long-range costs and are neglected in cost benefit analyses for supporting conservation practices.

Irrigation-Induced Erosion . . . A Separate Case

Despite the importance of irrigated agriculture to world food and fiber supplies, and its role in preserving other threatened natural ecosystems, irrigation-induced erosion has received very little attention in the mainstream erosion research community (Larson et al., 1990). Progress toward raising conservationists' and scientists' collective consciousness about the topic has been slow, but some important papers and reviews have appeared in recent years (Carter, 1990; Carter, 1993; Carter et al., 1993; Koluvek et al., 1993; Trout and Neibling, 1993).

Research on irrigation-induced erosion only began appearing with any frequency in the 1970s, following establishment of the Kimberly, Idaho, Agricultural Research Service research group in 1964. To date, scientists at the Kimberly location have published over 100 related papers. Because this group had the advantage of drawing conceptually from the extensive body of existing rain-induced research, development of irrigation-induced erosion research has proceeded somewhat holistically; that is to say, description, parameterization, theory and management research all occurred more nearly concurrently than was possible initially for rain and wind-induced erosion.

Irrigation-induced erosion has generally been treated by the unfamiliar as a rudimentary subset of rain-induced erosion systematics. Consequently, irrigation-induced erosion theory has been approached through a series of modifications of rain-induced erosion theory, often using experimental techniques and empirical relationships specifically and exclusively designed to simulate rain-fed processes (Trout and Neibling, 1993; Trout, 1998).

Irrigation water "encounters" soil in a variety of systematically different ways than rainfall, all of which are manageable to varying degrees. While the physical and chemical processes involved in rain-induced erosion and irrigation-induced erosion are the same, the systematics (application, chemistry, energy components, and mass balance) governing irrigation differ vastly from rainfall. For these reasons, both theory and management of irrigation-induced erosion differ significantly from rain-induced erosion theory and rain-fed agricultural management. These differences in systematics can be easily identified, although modification of existing rain-fed theory, management, and mind-set to accommodate these realities has proved somewhat daunting.

Rain-induced erosion theory essentially excludes consideration of water-soil interactive chemistry per se, dealing only with factors that describe stream and/or shower quantity and intensity. Rainwater quality is nearly uniform enough planet-wide (low electrical conductivity—EC and low sodium adsorption ratio—SAR) that variations in these parameters can be largely ignored without consequence in analyzing rain-induced erosion. Nonetheless, miniature laboratory simulations (Levy et al., 1994, Shainberg et al., 1994) and field studies (Le Bissonnais and Singer, 1993; Lentz et al., 1993, 1996) have demonstrated that water quality, especially EC and SAR do significantly impact the erosiveness of a given shower or stream of water.

Other specific physical and chemical components vary in natural and applied waters (e.g., temperature or dissolved organics) and may also have measurable effects. Soil and water temperatures vary systematically during the season and may partially explain some seasonal trends in erosion (Brown et al., 1995). Changes in aggregate stability have been attributed to small changes in soluble soil organic constituents (Coote et al., 1988; Harris et al., 1966; Young et al., 1990).

In existing models, specific soil chemistry-related factors are considered only inasmuch as they contribute to intrinsic soil erodibility, or are affected through various management-related parameters. The evolution to current process-driven erosion theories and models, e.g., the Water Erosion Prediction Project (WEPP) models (Laflen et al., 1991) have yet to meaningfully embrace interactions of soil and water chemistry, which in arid irrigated soils are very important to soil physical and hydraulic behavior and are spatially and temporally variable.

Irrigation water quality varies seasonally and geographically and can even vary on the farm, both between irrigation sets and within an irrigation set, if multiple water sources are involved (conjunctive water use). High-SAR/low-EC water is more erosive than Low-SAR/high-EC water (Figure 3). Irrigation water may or may not contain a substantial sediment or suspended biotic load upon entering a furrow. Those loads change systematically as the stream advances. Furrow erosion is affected by the initial sediment load of irrigation water due both to carrying capacity effects and surface sealing (Brown et al., 1988; Foster and Meyer, 1972).

Water advancing down a dry furrow instantaneously wets dry soil, destroying soil structure more pervasively than water collecting in wet rills during rainfall (Lentz et al., 1993, 1996). Aggregate stability can be affected by soil water content and gradual water content changes (Bullock et al., 1988; Kemper and Rosenau, 1984). Furrow stream size, which is exponentially related to detachment (Kemper et al., 1985), decreases down an irrigation furrow (because of cumulative infiltration effects), whereas wetted perimeter generally broadens from accumulated deposition. In rain-fed rills, soil is uniformly wet

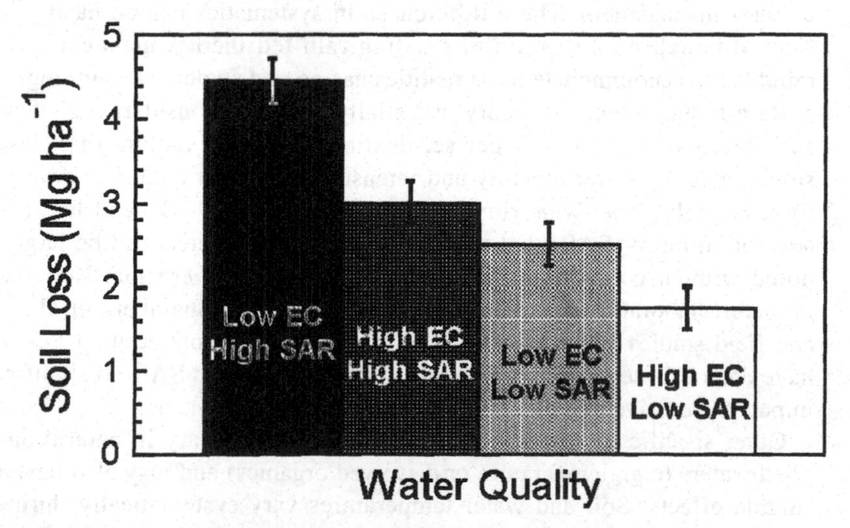

Figure 3. The effect of four water qualities on soil lost in tailwater from irrigation furrows. (Adapted from Lentz et al., 1993).

and runoff continuously increases down slope, with comparatively little deposition in the rill. Rill erosion, partitioned from rainfall simulator results, is not analogous to furrow erosion which, in addition to the considerations already mentioned, has no splash or water-drop component nor any intercepted interrill runoff or sediment.

Profile soil water content strongly influences erosion rate (Berg and Carter, 1980; Kemper et al., 1985). In furrow irrigation, it varies progressively along the irrigation path. In contrast, soils in level rain-fed agricultural landscapes generally have similar soil water profiles for most points on the landscape at any time in a rainfall event. Furthermore, irrigation-induced erosion results from a predictable series of nearly uniform and manageable water applications that must be accounted for in whatever conceptual model is used to estimate or predict erosion. Irrigation-induced erosion cannot be assessed by deriving yearly or seasonal relationships based on meteorological inputs averaged from sporadic events of varied intensity, occurring over long time periods across a geographic region. This obstacle is compounded if the amount and kind of irrigation is not accounted for.

Lehrsch and Brown (1995) were unable to correlate furrow irrigation-induced erosion with aggregate stability. Trout (1998) found that transport relationships used in WEPP did not predict erosion measured in the field, and that sediment transport was more sensitive to shear and flow rate than predicted in the field, possibly because shear did not vary as predicted. Furthermore, WEPP is a steady-state process-driven model and most types of irrigation are nonsteady-state processes. Kemper et al. (1985) noted that critical shear for furrow irrigation is essentially zero. Bjorneberg (personal communication, 1997) has found that an exponential function better describes the relationship between detachment capacity and hydraulic shear stress than the linear critical shear function currently used in the WEPP model. These examples represent problems with various modeling parameters derived from rain-fed systematics currently used in WEPP to describe erosion process and that fail to adequately match the observed relationships for irrigation-induced erosion. These failures reinforce doubts that adequate predictive capability derived from rain-fed observations can be devised for irrigation-induced erosion and emphasize the need to better characterize the specific phenomena and unique systematics involved in the processes driving irrigation-induced erosion.

Soil Conservation Practices for Irrigated Agriculture

Most of the initial impetus for soil conservation in irrigated agriculture was protection of riparian areas receiving irrigation return flows. This led to an early strategy focused mainly on sediment settling basins in return flow systems. Subsequently, efforts concentrated on prevention of soil loss from the farm. A parallel goal of both of these containment strategies was to return captured sediment to eroded sites on farm land. Current research emphasis represents a shift from engineering practices toward development of soil, water, and crop management practices aimed at halting all soil movement, thereby retaining soil in place, eliminating subsequent soil handling or transport.

Because each farm is unique, a given sediment containment practice may not be equally suited to all situations. Farmers determine which practice or practices suit their individual situation. Ultimately, erosion abatement practices that are used are more valuable than practices that are not used, regardless of the relative potential effectiveness of a given practice. Enforcement of clean water standards may eventually demand that return flows leaving a farm meet specified water quality standards. These standards may be voluntary standards or may be tied to potent financial incentives or disincentives. In 1997 the Idaho courts required establishment of Total Mean Daily Load (TMDL) limits for 962 water quality impaired stream segments, and gave five years for compliance. At the same time the Northside Canal Company, which manages irrigation diversion along the north shore of the mid-reach of the Snake River, changed its water contracts to allow it to terminate water delivery to farms that seriously impair water quality of return flows.

Below is a brief summary of the more important conservation practices already developed for irrigated agriculture. Practices differ in ease of adoption, effectiveness, and cost of implementation, but offer a range of options to suit most situations. These practices and related factors have been discussed in greater detail in several recent publications (Carter, 1990; Carter et al., 1993; Sojka and Carter, 1994).

Sediment Retention Basins

Sediment ponds can be large, perhaps 1/10th ha (1/4 acre), servicing a 16–24 ha (40–60 acre) field, or small "mini-basins" that temporarily pond runoff for only 6–12 furrows. The basins reduce flow rates and briefly retain water, allowing deposition of suspended particulates and reducing desorption of phosphorous. Retention basin effectiveness depends on sediment load, inflow rates, retention time, and texture of suspended particulates. About 2/3 of solids can be removed from return flows, but only about 1/3 of the suspended clay and total P (Brown et al., 1981). Clay, where most adsorbed P resides, is slow to sink to the pond floor. Thus, the practice is more effective for medium textured soils than for clayey soils.

Buried-Pipe Erosion Control Systems

Buried drainpipes with vertical inlet risers allow furrow irrigation tailwater to pond at the lower reaches of fields until the water level initiates drainage into the riser. These systems promote sediment retention much as ponds do, and are often an adjunct to mini-basins. The method is best suited to elimination of concave field ends. Effectiveness is near 90% while concavities or basins are filling, but drops to pond efficiencies once depressions are filled (Carter and Berg, 1983).

Vegetative Filter Strips

Cereal, grass, or alfalfa (*Medicago sativa*) strips 3–6 m (10–20 feet) wide sown along the lower ends of furrow-irrigated row crop fields reduce sediment in runoff 40–60%, provided furrows are not cut through the filter strip area (Carter et al., 1993). Harvested filter strips yield 30–50% below normal for the strip crop.

Twin Row and Close Row Planting

Planting corn (*Zea mays*) as close as possible to both sides of an irrigated furrow to form twin row spacings halved field sediment loss in two years of observation (Sojka et al., 1992). Results for single but narrower row spacings were more variable but showed promise for corn, sugar beet (*Beta vulgaris*) and field beans (*Phaseolus vulgaris*). Erosion reduction results from a combination of factors including soil binding by roots in close proximity to the flow, introduction of plant litter into the furrow stream, and (with narrow rows only) systematic increase in furrow numbers (and hence wetted perimeter), thus reducing irrigation set duration needed to deliver equivalent quantities of water. This reduces runoff stream size and runoff period relative to total inflow.

Tailwater Reuse

Retention ponds can be inexpensively enhanced to recirculate sediment-laden water into the furrow irrigation water supply. This does not halt or slow erosion per se, but largely automates replacement of sediment onto fields from which they came. Advantages include maximizing water supply efficiency and 100% on farm sediment retention (Carter et al., 1993). Capital and energy cost and accelerated pump wear are disadvantages. There is also mingling of disease inoculum, weed seed, and chemicals, although these occur where return flows are reused anyway. On a larger scale, however, many surface irrigation districts have been engineered and are operated with an assumption of return flows making part of the irrigation supply for large portions of the district. Elimination of all return flows could dry up some reaches of existing systems or require modification of primary canal capacity to provide water to farms on lower reaches of the delivery system if some water is not routed through return flow systems.

Improved Water Management

Improved inflow/outflow management, stream size monitoring (post-advance flow reduction), field leveling, alternate furrow irrigation, infiltration measurement (soil water budget monitoring), and irrigation scheduling (furrow or sprinkler), can all improve water use efficiencies. These changes could reduce water application and hence, runoff amounts, reducing erosion as a side benefit (Trout et al., 1994).

Furrow Mulching

Use of plant residue or living mulches in irrigation furrows can be very effective at halting erosion. Permanent furrow sodding halted nearly 100% of erosion (Cary, 1986) without adverse yield effects in barley (*Hordeum vulgare*), wheat (*Triticum aestivum*), beans, and corn. The technique required a special furrow cutter to maintain established furrows. Straw or other manageable residues can be selectively placed in furrows producing 52–71% sediment loss reduction (Miller et al., 1987; Aarstad and Miller, 1981; Brown, 1985; Brown and Kemper, 1987). Drawbacks of these techniques include large increases in advance times and infiltration rates, and the addition of field operations for establishment and/or maintenance of mulches. Mulching can occur at inconvenient times for crop managers, or cause problems during cultivation. Straw also sometimes moves in furrow streams, damming furrows and causing water to flow over rows into adjacent furrows.

Whey Application

Some irrigated areas are near dairy processing plants. For many processors, disposal of acid cottage cheese whey is a problem. Soil-applied acid whey accelerates remediation of exposed lime subsoils and conserves nutrients, using an agricultural by-product. If combined with straw application, whey can reduce furrow irrigation-induced erosion as much as 98% and increase infiltration over 20% (Robbins and Lehrsch, 1992; Brown and Robbins, 1997; Lehrsch and Robbins, 1994). The disadvantages of this approach are the cost and inconvenience of bulk hauling and field application of the whey. Usually processors, who often need land application sites, will provide whey at no cost.

Polyacrylamide-Treated Irrigation Water

Treating advancing furrow irrigation water (only) with 10 gm^{-3} polyacrylamide (PAM) has reduced sediment loss in runoff 85–99% while increasing infiltration 15 % (Lentz et al., 1992; Lentz and Sojka, 1994; Sojka and Lentz,

1993, 1994). This translates to about 1 kg ha^{-1} of PAM used per treated irrigation. PAM, an industrial flocculent used for food processing and water treatment, is now marketed extensively (over 162,000 ha treated in the USA in 1996) for erosion control in irrigated agriculture. Results have been highly consistent on a wide range of soils and conditions, showing high effectiveness, low cost, and lack of major effects on other farming practices (Figure 4). With 10 g m^{-3} PAM, initial water inflows can be more than doubled (then cut back once water has advanced across the field), virtually without erosion. This permits greater field infiltration uniformity. Ongoing PAM research by conservationists and manufacturers is rapidly providing better materials and more effective user protocols. Interest has also arisen for use of PAM with sprinkler irrigation.

Water Quality

In recent field research at Kimberly, ID, elevated SAR in furrow irrigation water, especially at low EC, increased the erosivity of the furrow stream (Lentz et al., 1996). Sediment in runoff more than doubled when SAR 12 EC

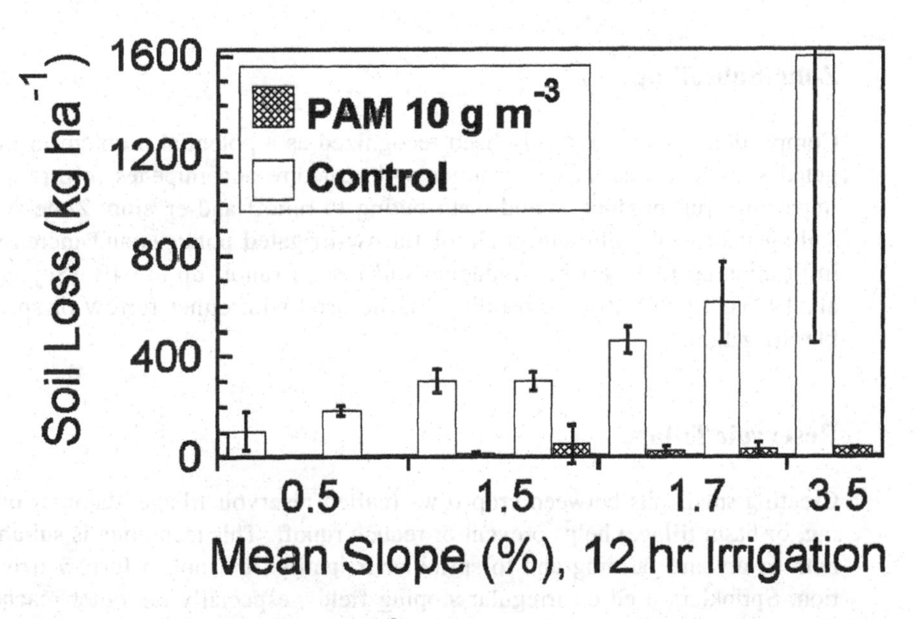

Figure 4. The effect of 10 gm^{-3} polyacrylamide in advancing furrow irrigation water (only) on soil lost in tailwater for the entire 12 hr duration of irrigation. (Adapted from Sojka and Lentz, 1994).

0.5 dS m^{-1} water was used, compared to SAR 0.7 EC 2.0 dS m^{-1} water. Sediment loss increased 1.5 times, compared to Snake River water (SAR 0.7 EC 0.5 dS m^{-1}). Many farms have multiple water sources (e.g., well and canal water) of varying quality. It behooves farmers to use less erosive water on steeper or more erosive ground, and/or to blend waters, where feasible, to reduce erosion hazard. These results further demonstrate that process-driven erosion models must consider water quality effects. They also underscore the need to know the quality of water used in erosion simulators for valid data interpretation.

Conservation Tillage

Field-wide erosion reductions of over 90%, reduced production costs, and some yield increases have been noted for a range of conservation tillage and no-till cropping systems under furrow irrigation (Carter and Berg, 1991; Sojka and Carter; 1994). Once established, these systems can provide long-range, cost-effective erosion elimination. A disadvantage of conservation tillage is a reluctance by many furrow irrigation farmers to adopt such all-encompassing changes to their operations. Furrow irrigation needs reasonably uniform and unobstructed furrows for consistent and timely water advance. This sometimes is a problem in residue-intensive systems. Under sprinkler irrigation, conservation tillage can be implemented much as in rain-fed systems.

Zone-Subsoiling

Compaction has only recently been recognized as a potential problem in irrigated soils. Compaction deteriorates soil structure and impedes infiltration, impairing crop production and contributing to runoff and erosion. Zone-subsoiling improved yield and grade of furrow-irrigated potatoes and increased infiltration up to 14% while reducing soil loss in runoff up to 64% (Sojka et al., 1993a, 1993b). Zone-subsoiling can be used with either furrow or sprinkler irrigation.

Reservoir Tillage

Creating small pits between crop rows (called reservoir tillage, dammer diking, or basin tillage) helps prevent or reduce runoff. This technique is suitable both to dryland farming and to sprinkler irrigation, but not to furrow irrigation. Sprinklers used on irregular sloping fields, especially the outer reaches of center pivots where application rates are high, can induce excessive runoff and erosion. Reservoir tillage has eliminated about 90% of these sprinkler-related runoff and erosion losses (Kincaid et al., 1990).

Low-Pressure Wide-Area Spray Emitters

The geometry of center pivot irrigation systems requires very high instantaneous water application rates in the outermost one-third of the pivot. The larger the pivot, the worse the problem. By using spray booms and special emitters, smaller drop sizes are spread over a larger area. Energy is conserved and runoff and erosion are greatly reduced compared to standard impact-head systems (Kincaid et al., 1990).

Conclusions

Despite great progress in the past 25 years, much work remains to achieve the needed understanding and control of irrigation-induced erosion. The importance of irrigation to feeding and clothing the world's growing population and protecting unspoiled habitats is not well appreciated by agriculturalists or the general public. Irrigation-induced erosion poses a significant threat to the sustainability of irrigated agriculture.

Most importantly, the uniqueness of irrigation-induced erosion as a phenomenon that is quite different from rain-fed erosion, is not well recognized by most of the erosion research community. These perspectives must be fostered in the erosion research community in order for research on irrigation-induced erosion and its abatement to be adequately prioritized and financed. The fundamental knowledge needed for development of theory and predictive capability for irrigation-induced erosion must advance, and conservation efforts for irrigated land should be strengthened.

The professional societies and their journals serving soil science, agriculture, and soil and water conservation have played key roles in the progress to date, and must continue to do so if these challenges are to be met. Priorities that deserve greater promotion include: (1) expanded public funding for the development of erosion theory and models that are derived specifically from and for irrigated systematics; (2) encouragement of cost/benefit analysis of government support for soil and water conservation programs and research efforts, based on relative productivity and risk to the land resources being husbanded; and (3) development of permanent efforts within the soil science community and its journals and among public policy makers to guarantee a balance of focus between rain-fed and irrigated agriculture, recognizing the crucial contribution of irrigation to meeting humanity's production needs while protecting the environment.

References

Aarstad, J.S., and D.E. Miller. 1981. Effects of small amounts of residue on furrow erosion. Soil Sci. Soc. Am. J. 45:116–118.

Anonymous. 1995. 1994 Irrigation survey. Irrig. J. 45:27–42.

Berg, R.D., and D.L. Carter. 1980. Furrow erosion and sediment losses on irrigated cropland. J. Soil Water Conserv. 35:367–370.

Brown, M.J. 1985. Effect of grain straw and furrow irrigation stream size on soil erosion and infiltration. J. Soil Water Conserv. 40:389–391.

Brown, M.J., J.A. Bondurant and C.E. Brockway. 1981. Ponding surface drainage water for sediment and phosphorous removal. Trans. ASAE. 24:1478–1481.

Brown, M.J., D.L. Carter, G.A. Lehrsch and R.E. Sojka. 1995. Seasonal trends in furrow irrigation erosion in southern Idaho. Soil Technology 8:119–126.

Brown, M.J., and W.D. Kemper. 1987. Using straw in steep furrows to reduce soil erosion and increase dry bean yields. J. Soil. Water Conserv. 42:187–191.

Brown, M.J., D.W. Kemper, T.J. Trout and A.S. Humpherys. 1988. Sediment, erosion and water intake in furrows. Irrig. Sci., 9:45–55.

Brown, M.J., C.W. Robbins and L.L. Freeborn. 1997 (anticipated). Combining cottage cheese whey and straw reduces erosion while increasing infiltration in furrow irrigation. J. Soil Water Conserv. (in press).

Bucks, D.A., T.W. Sammis, and G.L. Dickey. 1990. Irrigation for arid areas. pp. 499–548. In: Management of Farm Irrigation Systems. G.J. Hoffman, T.A. Howell, and K.H. Solomon (eds.). Am. Soc. Ag. Eng., St. Joseph, MI.

Bullock, M.S., W.D. Kemper and S.D. Nelson. 1988. Soil cohesion as affected by freezing, water content, time and tillage. Soil Sci. Soc. Am. J. 52:770–776.

Carter, D.L. 1990. Soil erosion on irrigated lands. In: Irrigation of Agricultural Crops, Ch. 37, Agronomy 30. pp. 1143–1171. B.A. Stewart and D.R. Nielson, eds., Am. Soc. Agronomy, Madison, WI.

Carter, D.L. 1993. Furrow erosion lowers soil productivity. J. Irrig. Drain. Eng. ASCE. 119:964–974.

Carter, D.L., and R.D. Berg. 1983. A buried pipe system for controlling erosion and sediment loss on irrigated land. Soil Sci. Soc. Am. J. 47:749–752.

Carter, D.L., and R.D. Berg. 1991. Crop sequences and conservation tillage to control irrigation furrow erosion and increase farmer income. J. Soil Water Conserv. 46:139–142.

Carter, D.L., C.E. Brockway, and K.K. Tanji. 1993. Controlling erosion and sediment loss from furrow-irrigated cropland. J. Irrig. Drain. Eng. ASCE. 119:975–988.

Cary, J.W. 1986. Irrigating row crops from sod furrows to reduce erosion. Soil Sci. Soc. Am. J. 50:1299–1302.

Coote, D.R., C.A. Malcolm-McGovern, G.J. Wall, W.T. Dickinson and R.P. Rudra. 1988. Seasonal variation of erodibility indices based on shear strength and aggregate stability in some Ontario soils. Can. J. Soil Sci. 68:405–416.

Fornstrom, K.J., and J. Borelli. 1984. Design and management procedures for minimizing erosion from furrow irrigated cropland. ASAE Pap. 84–2595. ASAE, St. Joseph, MI.

Foster, G.R., and L.D. Meyer. 1972. A closed-form soil erosion equation for upland areas. In: H.W. Shen (ed.) Sedimentation (Einstein). Colorado State Univ., Fort Collins, CO, pp. 12–1 to 12–19.

Gleick, P.H. (ed). 1993. Water in Crisis. Oxford University Press, New York.

Harris, R.F., G. Chesters and O.N. Allen. 1966. Dynamics of soil aggregation. Adv. Agron. 18:107–169.

Hoffman, G.J., T.A. Howell and K.H. Solomon (eds.) 1990. Management of Farm Irrigation Systems. ASAE, St. Joseph, MI.

Hunst, M.A., and B.V. Powers (eds.) 1993. Agricultural Statistics, 1993. U.S. Govt. Printing Office.

Kemper, W.D., and R.C. Rosenau. 1984. Soil cohesion as affected by time and water content. Soil Sci. Soc. Am. J. 48:1001–1106.

Kemper, W.D., T.J. Trout, M.J. Brown, and R.C. Rosenau. 1985. Furrow erosion and water and soil management. Trans. ASAE 28:1564–1572.

Kincaid, D.C., I. McCann, J.R. Busch and M. Hasheminia. 1990. Low pressure center pivot irrigation and reservoir tillage. pp. 54–60. In: Visions of the Future. Proc. 3rd Natl. Symp., Phoenix, AZ, Oct. 28–Nov. 1, 1990, ASAE, St. Joseph, MI.

Koluvek, P.K., K.K. Tanji and T.J. Trout. 1993. Overview of soil erosion from irrigation. J. Irr. Drain. Eng. ASCE. 119:929–946.

Laflen, J.M., J. Lane and G.R. Foster. 1991. The water erosion prediction project—a new generation of erosion prediction technology. J. Soil and Water Conserv. 46:34–38.

Larson, W.E., G.R. Foster, R.R. Allmaras and C.M. Smith (eds.). 1990. Proceedings of Soil Erosion and Productivity Workshop. University of Minnesota, St. Paul.

Le Bissonnais, Y., and M.J. Singer. 1993. Seal formation, runoff, and interrill erosion from seventeen California soils. Soil Sci. Soc. Am. J. 57:224–229.

Lehrsch, G.A., and M.J. Brown. 1995. Furrow erosion and aggregate stability variation in a Portneuf silt loam. Soil Technology 7:327–341.

Lehrsch, G.A., and C.W. Robbins. 1994. Cheese whey as an amendment to disturbed lands: Effects on soil hydraulic properties. In: Proc. Int. Land Reclamation and Mine Drainage Conf. and Third Int. Conf. on the Abatement of Acidic Drainage, Pittsburgh, PA. Apr. 1994. Vol. 3. USDI Bur. Mines Spec. Publ. SP 06C-94. NTIS, Springfield, VA.

Lentz, R.D., I. Shainberg, R.E. Sojka, and D.L. Carter. 1992. Preventing irrigation furrow erosion with small applications of polymers. Soil Sci. Soc. Am. J. 56:1926–1932.

Lentz, R.D., R.E. Sojka and D.L. Carter. 1993. Influence of irrigation water quality on sediment loss from furrows. Proceedings Vol. II of the SWCS Conference on Agricultural Research to Protect Water Quality, 21–24 Feb. 1993, Minneapolis, MN, pp. 274–278.

Lentz, R.D. and R.E. Sojka. 1994. Field results using polyacrylamide to manage furrow erosion and infiltration. Soil Science. 158:274–282.

Lentz, R.D., R.E. Sojka and D.L. Carter. 1996. Furrow irrigation water-quality effects on soil loss and infiltration. Soil Sci. Soc. Am. J. 60:238–245.

Levy, G.J., J. Levin and I. Shainberg. 1994. Seal formation and interill soil formation. Soil Sci. Soc. Am. J. 58:203–209.

Mech, S.J. 1959. Soil erosion and its control under furrow irrigation in arid West. Info Bull. 184. USDA/ARS Gov. Print. Office. Washington, DC.

Miller, D.E., and J.S. Aarstad and R.G. Evans. 1987. Control of furrow erosion with crop residues and surge flow irrigation. Agron. J. 51:421–425.

Robbins, C.W., and G.A. Lehrsch. 1992. Effects of acidic cottage cheese whey on chemical and physical properties of a sodic soil. Arid Soil Res. and Rehab. 6:127–134.

Ross, M.A., and C.A. Lembi. 1985. Applied Weed Science. Macmillan and Company, New York.

Shainberg, I., J.M. Laflen, J.M. Bradford and L.D. Norton. 1994. Hydraulic flow and water quality characteristics in rill erosion. Soil Sci. Soc. Am. J. 58:1007–1012.

Sojka, R.E., M.J. Brown and E.C. Kennedy-Ketcheson. 1992. Reducing erosion from surface irrigation by furrow spacing and plant position. Agron. J. 84:668–675.

Sojka, R.E. and D.L. Carter. 1994. Constraints on Conservation Tillage Under Dryland and Irrigated Agriculture in the United States Pacific Northwest. Ch. 12. In: M.R. Carter (ed.) Conservation Tillage in Temperate Agroecosystems. Lewis Publishers, Boca Raton, FL.

Sojka, R.E., and R.D. Lentz. 1993. Improving water quality of return flows in furrow-irrigated systems using polymer-amended inflows. Proceedings of the SWCS Conference on Agricultural Research to Protect Water Quality, 21–24 Feb., 1993. Minneapolis, MN, pp. 395–397.

Sojka, R.E., and R.D. Lentz. 1994. Polyacrylamide (PAM): A new weapon in the fight against irrigation-induced erosion. USDA-ARS Soil and Water Management Research Unit, Station Note #01-94.

Sojka, R.E., and R.D. Lentz. 1995. Decreasing sedimentation loss from surface irrigation into riparian areas. Proceedings: Northwest Regional Riparian Symposia. Diverse Values: Seeking Common Ground. Idaho Riparian Cooperative. University of Idaho and the Idaho Water Resources Research Institute. 8–9 December, 1994. Boise, ID, pp. 62–68.

Sojka, R.E., D.T. Westermann, M.J. Brown and B.D. Meek. 1993a. Zone-subsoiling effects on infiltration, runoff, erosion, and yields of furrow-irrigated potatoes. Soil & Till. Res. 25:351–368.

Sojka, R.E., D.T. Westermann, D.C. Kincaid, I.R. McCann, J.L. Halderson, and M. Thornton. 1993b. Zone-subsoiling effects on potato yield and grade. Am. Pot. J. 70:475–484.

Sojka, R.E. 1997. Irrigation. In: Sybil Parker (ed.) 1997 Yearbook of Science and Technology. pp. 258–260. McGraw-Hill, New York.

Stevenson, F.J. 1972. Organic matter reactions involving herbicides in soils. J. Environ. Qual. 1:333–343.

Tribe, D. 1994. Feeding and Greening the World, the Role of Agricultural Research. CAB International. Wallingford, Oxon, United Kingdom.

Trout, T., D. Carter, and B. Sojka. 1994. Irrigation-induced soil erosion reduces yields and muddies rivers. Irrigation J. 44:8, 11–12.

Trout, T.J., and W.H. Neibling. 1993. Erosion and sedimentation processes on irrigated fields. J. Irr. Drain. Eng. ASCE. 119:947–963.

Trout, T.J. 1996. Furrow irrigation erosion and sedimentation: on-field distribution. Trans. ASAE 39:1717–1723.

Trout, T.J. 1998 (Anticipated). Modelling sediment transport in irrigation furrows. Soil Sci. Soc. Am. J. (Submitted).

Young, R.A., M.J.M. Romkens, and D.K. McCool. 1990. Temporal variations in soil erodibility. In: R.B. Bryan (ed.) Soil Erosion—Experimental Models. Catena Suppl. 17. Catena Verlag, Lawrence, KS, pp. 41–53.

Mechanics, Modeling, and Controlling Soil Erosion by Wind

D.W. Fryrear and J.D. Bilbro

Introduction

Wind erosion, or at least the dust and sand drifts associated with wind erosion, have presented a problem to mankind since biblical times. Severe dust storms were reported in Chinese history over 900 years ago. Wind erosion has been a factor in shaping the earth's surface. Geomorphological landscape features reflect that tremendous quantities of loess soils have been deposited by wind over hundreds or thousands of years. Portions of Central Asia, Middle East, and North Africa supported prosperous populations, but improper land use and excessive erosion have resulted in their present barren state (Chepil and Woodruff, 1963).

In other regions, once-fertile soils have been rendered infertile by the complete removal of topsoil (Fryrear, 1990). Red eyes (Lewis and Clark, 1806), asthma or respiratory problems (Chow, 1995), traffic accidents, and lower crop yields (Lyles, 1977) are a few of the additional problems associated with wind erosion. Dust was a part of the environment long before the Great Plains was plowed (Malin, 1946).

The following quote made after a tour of the drought-ravaged Great Plains illustrates the impact of severe and continuous wind erosion on people.

> Wind erosion in the United States has developed to the point where it has become a very serious national problem. It is serious not only from the standpoint of tremendous damage already done but also from the standpoint of probable future damage of much greater proportions, unless effective control measures are soon applied. In addition to the very serious damage to the soils, there has also been widespread and serious damage to people, crops, livestock, native pastures, roads, automobiles, buildings, goods on the shelves of mercantile establishments, farm machinery, and many other things. Far too few people realize the present gravity of the situation, and probably no one appreciates the full significance of the threat to future generations. (Joel, 1937).

The role of wind erosion research has been to identify mechanics of the process, design effective control practices, and develop models that estimate soil erosion by wind. The milestones of research and their impact on wind erosion include, but are not limited, to the following:

1. wind erosion mechanics
2. field wind erosion samplers
3. wind erosion models
4. multiple control practices.

Wind Erosion Mechanics

Basic research on the mechanics of wind was conducted by Bagnold (1943), Chepil and Milne (1939), and Zingg, Chepil, and Woodruff (1953) to identify the major factors contributing to erosion. These basic factors include wind, soil surface roughness, and vegetation conditions.

Surface Aerodynamic Properties 'k,' 'Zo,' and 'D'

When wind velocity is plotted against the log of height above the mean aerodynamic surface, the result is a straight line intersecting the height line (zero velocity) at **"k"**. The "k" has no relationship to the height of vegetation or other soil roughness elements, but for rough surfaces there is momentum transfer across the elements to the stable surface. Contrary to previous reports (Chepil and Woodruff, 1963), the k varies with surface conditions and wind velocity. This variation was measured in the absence of soil movement, but k is more variable when soil erosion is occurring.

Zo—The influence of roughness elements is reflected in the aerodynamic roughness. The rougher the surface the larger the eddies of wind gust; therefore, Zo represents the eddy size at the surface (Panofsky and Dutton, 1984). As wind blows over a rough soil surface, the wind profile is displaced to a new reference plane.

D—The degree to which the wind profile is displaced is called the **displacement height** (Lowry, 1967). The displacement height is also defined as the depth of still air trapped among the roughness elements (Sutton, 1949), and may be called roughness height (Geiger, 1957). **Total surface roughness** may be determined by adding the aerodynamic surface roughness and displacement height (Chepil and Woodruff, 1963).

Drag Velocity

As the wind blows over a surface, the surface exerts a drag on the wind. The rougher the surface, or the higher the wind velocity, the greater the drag. As the drag increases, soil particles may be dislodged and injected into the wind stream. Because wind velocity increases with height above the soil surface, airborne soil particles also gain velocity as they gain altitude. The wind continues to exert a drag on the particles. If the particles are small they may become suspended in the wind stream and transported great distances. Many particles are too large to be suspended, and these particles continue to gain velocity as they return to and strike the soil surface in a transport mode called saltation. The saltating particles abrade soil or plant surfaces.

Threshold Velocity

The minimum wind velocity at which the most unstable particles are dislodged is called the threshold velocity. The static threshold velocity (also called **minimal fluid threshold velocity,** Chepil and Woodruff, 1963) is the velocity at which the most unstable particles are dislodged, but soil movement is not sustained. The dislodged particle may be trapped in the wake of larger particles, or may dislodge additional particles. If the wind velocity increases, additional unstable particles move and this process continues until soil particle movement is sustained without a further increase in velocity. The velocity at which soil movement is sustained is called the dynamic threshold velocity (also called **impact threshold velocity** Bagnold, 1943). The calculation of "threshold velocity" is not simple, but must include surface soil moisture content, soil roughness, residue levels, standing residues, and wind turbulence.

Wind Profiles

Wind is the basic driving force in wind erosion. Wind velocities are never constant and are characterized by eddies moving at variable velocities in all directions (Chepil and Woodruff, 1963). A wind strong enough to erode soil is always turbulent. The profile of the wind can be described using Equation 1 (Brunt, 1944; von Karman, 1934)

$$Vz = 5.75\ V^{*} \log z/k \tag{1}$$

where

Vz = velocity at any height z,
V* = drag velocity,
z = height above the soil surface,
k = index of aerodynamic surface roughness.

While this relationship does not hold above the surface boundary layer or when the surface is eroding, Equation 1 is applicable close to the soil surface where the wind is fully turbulent.

Transport Rate

The quantity of material being transported will increase as the wind velocity cubed. Any control method that reduces the velocity will greatly reduce the transport rate. As the quantity of soil being transported by wind increases, the velocity of the wind is reduced at the soil surface (Bagnold, 1943; Chepil and Milne, 1939). The end result is that as the velocity of the wind increases, the velocity close to the soil surface may remain constant or decrease slightly due to the greater concentration of saltating soil grains, and the higher rate of movement (Chepil, 1945).

For eroding soil surfaces the description of drag velocity may not be well defined. Instead of drag velocity, wind velocity at a standard height can be used. The coefficients may also vary with size of erodible particles, concentration of dust in the airstream, and the proportion and size of nonerodible aggregates.

Soil

Wind may be the driving force in wind erosion, but soil and condition of the soil surface will impact the rate at which wind can detach and transport soil particles. Because of the weak bonding between particles, sandy soils are more susceptible to erosion by wind than fine textured soils. However, any soil may be eroded by wind if the surface is dry, finely divided, and not protected with nonerodible material.

Erodible Fraction

All soils contain nonerodible and erodible material. For most field conditions, any particle smaller than 0.84 mm is considered erodible. The erodible fraction can be determined from a standard wind tunnel test or from a compact rotary sieve developed to determine size distribution of surface soils (Chepil, 1962). The rotary sieve permits the quantification of erodible fraction from field plot tests or from samples collected and shipped hundreds or thousands of miles (Fryrear et al., 1994).

Soil Roughness

If the soil surface is roughened with tillage, erosion may be controlled until aggregates on the soil surface are reconsolidated with subsequent rainfall. For coarse textured soils, rainfall will quickly smooth the surface by disintegrating soil clods. The resulting soil surface may be easily eroded by wind. Fine-textured soil aggregates will not disperse as quickly and the resulting soil surface may be less erodible by wind.

A smooth soil surface is much more erodible than a rough cloddy soil with erodible particles between soil clods. Ridging cultivated soils may be an effective and efficient method of reducing soil erosion (Fryrear, 1984). If the soil ridges are composed of erodible and nonerodible aggregates and the surface begins to erode, the eroded material will move from the ridge and be trapped in the furrows. The ridge will soon be stabilized with a mantle of nonerodible soil aggregates (Chepil and Woodruff,1963).

Vegetation Characteristics

Managing plant residues is one of the major methods of controlling wind erosion, but not all crops produce sufficient quantities of residue to protect soils from wind erosion (Bilbro and Fryrear, 1985).

Flat Residue

When crops are harvested and the fields tilled to control weeds, the residues are usually flat on the soil surface. A 30% cover on the soil surface will reduce soil erosion by 70% (Fryrear, 1985). When expressed on the basis of the percent of the soil surface that is covered by the flat material, the type of residue vegetation (diameter, density, or type) does not significantly affect the relationship between soil cover and wind erosion.

Standing Residue

Crop residues standing in the field are six times as effective in controlling wind erosion than when the same quantity of material is flat on the soil surface (Bilbro and Fryrear, 1994). Several types, sizes, and heights of simulated vegetation have been tested and the relationships combined into a single expression (van de Ven et al., 1989).

Soils that are covered with growing vegetation are not easily eroded by wind. Growing a vegetative cover is an effective erosion control practice in regions with abundant rainfall. Sufficient water to grow a cover crop is not always available in many wind erosion problem areas. When water is limiting, residues (both flat and standing) from previous crops must be maintained on the soil surface. If residues are insufficient to protect the soil, roughening the soil surface may be a viable option.

Wind Erosion Sampler (BSNE)

Between the publication of the Wind Erosion Equation (WEQ) (Woodruff and Siddoway, 1965) and the mid-1980s, numerous laboratory and field wind tunnel studies were conducted. The conversion of laboratory wind tunnel results to field conditions was always a concern because field verification of the laboratory results was not possible. Measurements of soil erosion in the field became possible with the development of the Big Spring Number Eight (BSNE) field wind erosion sampler. The BSNE has been used extensively to collect field samples of soil being eroded by natural winds (Fryrear, 1986; Larney et al., 1995). This sampler also provided the first field erosion samples to determine erosion quantity, vertical distribution, and the effect of field length on soil movement by wind.

Wind Erosion Models

The WEQ introduced a new era of wind erosion prediction. For the first time an equation was available to aid in the planning of wind erosion control practices. With WEQ, erosion is a *function* of the soil erodibility, soil roughness, climate, field length, and vegetation.

The WEQ has been used throughout the United States of America, and the basic equation has been adopted by the Food and Agriculture Organization (FAO, 1979). The WEQ has been used to evaluate the status of soil resources by the Natural Resouces Conservation Service (NRCS) from the five-year program called NRI (National Resource Inventory).

Current programs require more precise and flexible erosion prediction technology. To address these concerns, the USDA Agricultural Research Service concentrated their scientific efforts on the development of the Revised Wind Erosion Equation (RWEQ) and the Wind Erosion Prediction System (WEPS). These wind erosion models incorporate new technology and represent the "state-of-the-art" in wind erosion technology.

Revised Wind Erosion Equation (RWEQ)

With the passage of the Food Security Act of 1985, it was imperative that cropping systems control wind erosion. The consensus of the wind erosion research scientists was that WEQ could not be modified to provide acceptable estimates of wind erosion, and new models that utilized advancements in wind erosion since the WEQ was published would improve estimates.

The RWEQ represents a completely different approach to modeling wind erosion (Core Group, 1994). In RWEQ the wind is the basic driving force of erosion and the transport capacity of the wind limits the movement of soil. In addition, the texture and roughness of the soil, the presence and condition of

Table 1. Comparision of Estimated Soil Losses with RWEQ 97 and Measured Soil Losses

	Field		Soil Loss	
Site	Shape	Size *ha*	Measured *kg/m²*	Estimated *kg/m²*
Canada #1*	circle	3.1	15.20	17.00
Canada #2*	circle	3.1	6.90	5.70
Akron, CO (88–89)	circle	2.5	0.83	6.59
Akron, CO (89–90)	circle	2.5	1.10	0.38
Eads, CO (90–91)	circle	2.5	2.43	2.76
Eads, CO (91–92)	circle	2.5	0.67	0.61
Crown Point, IN (90)	circle	2.5	31.21	25.24
Crown Point, IN (91)	circle	2.5	23.95	14.52
Crown Point, IN (92)	circle	2.5	0.42	1.55
Elkhart, KS (90)	circle	2.5	0.29	2.53
Elkhart, KS (91)	circle	2.5	1.32	9.48
Elkhart, KS (92)	circle	2.5	15.50	21.11
Elkhart, KS (93)	circle	2.5	2.06	13.87
Crookston, MN (88–89)	circle	2.5	0.21	0.00
Crookston, MN (89–90)	circle	2.5	0.32	0.20
Swan Lake, MN (91)	circle	2.5	1.14	1.55
Swan Lake, MN (92)	circle	2.5	0.13	0.00
Swan Lake, MN (93)	circle	2.5	0.00	0.00
Kennett, MO (92–93)	circle	2.5	13.73	6.86
Kennett, MO(93–94)	circle	2.5	0.64	1.66
Havre, MT (92–93)	circle	2.5	0.01	2.69
Havre, MT (93–94)	circle	2.5	0.01	0.63
Lindsey, MT (90–91)	circle	2.5	0.03	2.98
Lindsey, MT (91–92)	circle	2.5	0.09	3.54
Scobey, MT (88–89)	circle	2.5	4.68	7.46
Scobey MT Fallow (89–90)	circle	2.5	1.34	0.16
Scobey, MT Stubble (89–90)	circle	2.5	0.39	0.00
Sidney, NE (88–89)	circle	2.5	0.52	0.67
Sidney, NE (89–90)	circle	2.5	0.38	0.07
Sidney, NE (90–91)	circle	2.5	2.29	5.65
Portales, NM (94–95)	circle	59	0.01	0.13
Fargo, ND (94–95)	circle	2.5	0.00	0.00
Fargo, ND (95–96)	circle	2.5	0.00	0.00
Fargo, ND (96–97)	circle	2.5	0.00	0.00
Big Spring, TX (89)	circle	2.5	21.54	23.73
Big Spring, TX (90)	circle	2.5	20.96	18.60
Big Spring, TX (91)	circle	2.5	0.10	2.53
Big Spring, TX (93)	circle	2.5	28.78	28.15
Big Spring, TX (94)	circle	2.5	17.16	11.47
Big Spring, TX (95)	circle	2.5	26.29	38.23
Big Spring, TX (96)	circle	2.5	3.99	9.64
Big Spring, TX (97)	circle	2.5	13.63	12.82
Martin-C, TX #2 (95)	rectangle	55	0.30	0.63
Martin-C, TX #3 (95)	rectangle	41	0.80	1.01
Martin-C, TX #4 (95)	rectangle	145	0.30	0.92
Plains E, TX (94–95)	rectangle	72	2.20	0.54
Plains E, TX (95–96)	rectangle	72	3.83	2.24
Plains B, TX (94–95)	rectangle	145	1.60	1.39
Plains B, TX (800m) (95–96)	rectangle	72	2.02	0.45
Plains B, TX (1600m) (95–96)	rectangle	145	1.55	0.54
Mabton, WA (90–91)	circle	2.5	3.68	3.85
Prosser, WA #1 (91–92)	circle	2.5	0.17	0.25
Prosser, WA #2 (92–93)	circle	2.5	0.32	1.30

*Measured erosion and RWEQ 5.01 estimated erosion supplied by Frank Larney, Lethbridge, Canada.

surface residue, wind barriers, and the management system being used limit the quantity of soil that can be transported by wind.

The RWEQ has been revised and tested to ensure that estimates of wind erosion losses agree with measured values for a variety of field, climatic, and management conditions (Table 1). Additional field data that represent more diverse field conditions are being collected to validate RWEQ estimates for more extreme conditions.

Wind Erosion Prediction System (WEPS)

The WEPS uses more process-oriented approaches to improve wind erosion prediction technology. WEPS contains submodels for CROP GROWTH, DECOMPOSITION, SOIL, HYDROLOGY, WEATHER, EROSION, and TILLAGE. WEPS is designed to contain process-based technology, but it is a much more complex model than WEQ or RWEQ and requires additional input parameters (Hagen, 1991).

As soils are eroded by wind, fertile soil components essential to the production of crops are lost. The on-site impacts of wind erosion are damaged or destroyed crop seedlings, reduced crop quality, and increased production expenses. The off-site impacts include reduced visibility, increased equipment maintenance, abrasion of paints or exposed surfaces, and aggravated public health (Chow, 1995). Dust from wind erosion is readily visible and attracts the attention of the general populace in the source areas (Fryrear et al., 1996). In dust transport regions, the impacts on the environment are being quantified by researchers.

Multiple Control Practices

Even though wind erosion is one of the geomorphological processes responsible for land forms across the country, the magnitude of potential erosion can be reduced when the available resources are properly utilized. The resources available to reduce wind erosion include:

1. surface residues
2. soil roughness from tillage
3. cover crops
4. reducing field length by proper orientation against prevailing wind direction
5. wind barriers or shelterbelts
6. strip cropping
7. chemical soil stabilizers.

The management of surface residues is one method of reducing wind erosion, but residue management requires that high residue producing crops are in the crop rotation. In most areas, residues cannot be maintained on the soil surface for more than 12 to 18 months. When residues are maintained erect they are several times more effective in reducing wind erosion than when the same quantity is lying on the soil surface. With the RWEQ, it is possible to evaluate when residue levels are insufficient to control erosion and must be supplemented with other practices. Acceptable practices depend on resources available, but include roughening the soil surface, addition of annual or perennial wind barriers, strip-cropping, and in severe conditions, chemical soil stabilizers to provide the level of protection needed. With RWEQ it is possible to identify critical erosion periods and then test alternative methods of reducing wind erosion.

Under more hazardous climatic conditions, the alternative methods may include several of the options. During more favorable conditions, a single control method may be sufficient.

Summary

Research has expanded our understanding of the physics of wind erosion. With the development of efficient and versatile erosion samplers, it was possible to verify laboratory wind technology under field conditions. With an understanding of wind erosion mechanics and the measurement of field erosion, it was possible to develop and validate new wind erosion models.

These new models will enable users (a) to more accurately evaluate the effectiveness of present wind erosion control practices and (b) to design new control systems that will more effectively reduce soil erosion by utilizing available resources. This new technology enables the development of control systems that most effectively utilize available resources.

References

Bagnold, R.A. 1943. The physics of blown sand and desert dunes. William Morrow and Co., New York. pp. 256.

Bilbro, J.D., and D.W. Fryrear. 1985. Effectiveness of residues from six crops for reducing wind erosion in a semiarid region. J. Soil Water Conserv. 40(4):358–360.

Bilbro, J.D. and D.W. Fryrear. 1994. Wind erosion losses as related to plant silhouette and soil cover. Agron. J. 86:550–553.

Brunt, D. 1944. Physical and dynamical meteorology, 3rd ed. Cambridge University Press, London and New York.

Chepil, W.S., 1945. Dynamics of wind erosion: I. Nature of movement of soil by wind. Soil Sci. 60:305–320.

Chepil, W.S. 1962. A compact rotary sieve and the importance of dry sieving in physical soil analysis. Soil Sci. Soc. Am. Proc. 26(1):4–6.

Chepil, W.S., and R.A. Milne. 1939. Comparative study of soil drifting in the field and in a wind tunnel. Sci. Agric. pp. 249–257.

Chepil, W.S., and N.P. Woodruff. 1963. The physics of wind erosion and its control. Adv. Agron. 15:211–302.

Chow, J.C. 1995. A&WMA's 1995 critical review: Measurement methods to determine compliance with ambient air quality standards for suspended particles. J. Air Waste Manage. Assoc. 45:320–382.

Core Group of Scientists and Conservationists. 1994. Revised Wind Erosion Equation (RWEQ). Published in the Current and Emerging Erosion Prediction Technology: Extended Abstracts from the Symposium. Published by the Soil and Water Conservation Society, August 10–11, pp. 46–48.

FAO, 1979. A provisional methodology for soil degradation assessment. Food and Agriculture Organization of the United Nations, Rome.

Fryrear, D.W. 1984. Soil ridges-clods and wind erosion. Trans. ASAE 27(2):445–448.

Fryrear, D.W. 1985. Soil cover and wind erosion. Trans. ASAE 28:781–784.

Fryrear, D.W. 1986. A field dust sampler. J.Soil Water Conserv. 41:185–199.

Fryrear, D.W. 1990. Wind erosion: Mechanics, Prediction, and Control. Springer-Verlag, New York. Advan. Soil Sci. 13:187–199.

Fryrear, D.W., C.A. Krammes, D.L. Williamson, and T.M. Zobeck. 1994. Computing the wind erodible fraction of soils. J. Soil Water Conserv. 49(2):183–188.

Fryrear, D.W., J.B. Xiao, and W.Chen. 1996. Wind erosion and dust. International Conference on Air Pollution from Agricultural Operations. Kansas City, MO. February 7–9, pp. 57–64.

Geiger, R. 1957. The climate near the ground. 2nd rev. Harvard University Press, Cambridge, MA.

Hagen, L.J. 1991. A wind erosion prediction system to meet user needs. J. Soil Water Conserv. 46(2):105–111.

Joel, A.H. 1937. Soil conservation reconnaissance survey of the Southern Great Plains wind-erosion area. USDA Tech Bul. 566.

Larney, F.J., M.S. Bullock, S.M. McGinn, and D.W. Fryrear. 1995. Quantifying wind erosion on summer fallow in Southen Alberta. J. Soil Water Conserv. 50(1):91–95.

Lewis and Clark Expedition, Letter Report sent to President Jefferson from Wallula, WA., April 27, 1806.

Lowry, W.P. 1967. Weather and Life. Academic Press. New York. pp. 199.

Lyles, L. 1977. Wind Erosion: Processes and Effect on Soil Productivity. Trans. ASAE 20(5):889–887.

Malin, J.C. 1946. Dust storms Part 1: 1850–1861, Kansas Historical Quarterly. 14(2):129–144.

Panofsky, H.A., and J.A. Dutton. 1984. Atmospheric Turbulence. Models and Methods for Engineering Applications. John Wiley and Sons. New York. p. 120.

Sutton, O.G. 1949. Atmospheric Turbulence. Nethuen and Company Ltd., London, p. 56.

van de Ven, T.A.M., D.W. Fryrear, and W.P. Spaan. 1989. Vegetation characteristics and soil losses by wind. J. Soil Water Conserv. 44(4):347–349.

von Karman, T. 1934. Turbulence and skin friction. J. Aeronaut. Sci. 1:1–20.

Woodruff, N.P., and F.H. Siddoway. 1965. A wind erosion equation. Soil Sci. Soc. Am. Proc. 29:602–608.

Zingg, A.W., W.S. Chepil, and N.P. Woodruff. 1953. Analysis of wind erosion phenomena in Roosevelt and Curry Counties, NM. Published by Region VI SCS, Albuquerque, NM. M-436.

Conservation Tillage for Erosion Control and Soil Quality

R.L. Blevins, R. Lal, J.W. Doran, G.W. Langdale, and W.W. Frye

Introduction

Conservation tillage has been called the greatest soil conservation practice of the 20th century. Conservation tillage reduces the intensity of tillage operations and allows farmers to manage crop residues on or near the soil surface (Carter, 1993). When residues are left on the soil surface, the mulch instead of the soil absorbs the energy of raindrops. This prevents the soil aggregates beneath the mulch from being broken apart and dispersed, and the soil is less prone to seal over and form a crust. Residue left at the surface usually slows the movement of water over the soil surface, allowing for more water infiltration (Edwards, 1995).

So many scientists, educators, and farmers have contributed to the development of conservation tillage that it would be impossible to properly recognize all their achievements here. Below we have briefly reviewed the development of conservation tillage research from its inception in the 1930s to the present day value of the practice to farmers striving to comply with the soil conservation requirements of the Food Security Act.

Evolution of Conservation Tillage

The current technology on conservation tillage is a result of 50 or more years of research and testing. We must go back to at least the 1930s when early versions of chisel-plowing were being tested in the Great Plains to find the beginning of modern conservation tillage. Early attempts in the 1930s to till the soil while leaving most of the crop residues included the efforts of a Georgia farmer, J. Mack Gowdy (Middleton, 1952), who designed the Bull-Tongue scooter (Langdale, 1994). At about the same time, Fred Hoeme and C.S. Noble, two innovative stubble-mulch farmers from the Great Plains began designing and building tillage implements that could penetrate the dry soils of the plains and still leave a mulch on the soil surface (Allen and Fenster, 1986).

The stubble-mulch concept, a forerunner of conservation tillage, was given a further boost in 1938 when J.C. Russel and F.L. Duley combined their efforts as a research team at Lincoln, Nebraska. Their classic research on soil, water, and residue management (Duley and Russell, 1939) made significant contributions to understanding the principles related to soil and water conservation.

During this same time, another research team (T.C. Peele and O.W. Beale) in South Carolina conducted mulch tillage research under southeastern USA conditions (Peele, 1942). These research efforts were coordinated and supported by M.L. Nichols, Research Chief of the USDA Soil Conservation Service. During the 1940s, agricultural engineers G.W. Nutt and W.N. Adams modified the Noble sweep plow to make it more effective for use on Southern Piedmont soils (Nutt et al., 1943).

Other early conservation tillage researchers included a team from Ohio State University, C.O. Reed, L.D. Baver, and C.J. Willard, that initiated a milestone study in 1937 to educate different tillage systems for corn production (Hill et al., 1994a). Mulch tillage was one of the six treatments evaluated. Results were poor for the mulch tillage treatment because weed control was inadequate. During the 1940s, E.H. Faulkner wrote a popular book entitled "Plowman's Folly." He described the damaging effect of using the moldboard plow and claimed that "no one has ever advanced a scientific reason for plowing." This was questioned by both farmers and researchers because alternatives to plowing at that time would not allow farmers to control weeds or plant into the residue. The organization of the Soil Conservation Society of America (now the Soil and Water Conservation Society) in 1945 further enhanced awareness of the need for conservation tillage research.

Purdue University agricultural engineer, Russel Poyner, and agronomist, George Scarseth, developed the M-21 till-planter in 1946, which allowed direct seeding into mulch (Hill et al., 1994a). This forerunner for conservation tillage planters tilled a 25-cm band for seed placement. During the 1940s, the herbicide 2,4-D [(2,4-dichlorophenoxy) acetic acid] was made available to farmers in the USA.

Conservation tillage research in the 1950s shifted from stubble mulch to "plow-plant" and "wheel-track" methods, particularly in the eastern portion of the USA (McAdams and Beale, 1959; Larson and Beale, 1961). These reduced tillage systems incorporated residues but left the surface rough to help control soil erosion. During this era, another plow-plant system was developed by J.C. McAllister (1962) called the lister tillage method. This practice opened a furrow for planting and covered most of the interrow area with soil from the furrow (Beale and Langdale, 1964).

Several breakthroughs in the late 1950s and early 1960s paved the way for successful conservation tillage, including no-tillage. The availability of atrazine [6-chloro-*N*-ethyl-*N*'-(1-methylethyl)-1,3,5-triazine-2,4 diamine], a residual herbicide for corn, and paraquat, an effective burndown herbicide, were major breakthroughs that allowed no-tillage to become practical. The other requisite development was the production of a commercially available no-tillage planter, the Allis-Chalmer AC 600 series, that used fluted (wavy) coulters to loosen a narrow band of soil in which the seed could be placed via double-disk opener planting attachments.

Field research in the 1950s by M.A. Sprague in New Jersey showed that chemicals could be used for weed control in pasture renovation studies. Other supporting conservation tillage research during this decade included G.C. Klingham's experimental studies on growing grain crops without tillage in North Carolina.

No-tillage studies for field crops were initiated in 1961 in Virginia (Moody et al., 1961) using atrazine and black plastic. Moschler et al. (1969) made significant contributions to maintaining better surface mulch by planting no-tillage corn into cover crops. Researchers in Texas began to test triazine herbicides for weed control (Wiese et al., 1967) in dryland environments. Ohio researchers (Triplett et al., 1964) in 1962 started long-term no-tillage plots on contrasting soil types. Purdue researchers, D.R. Griffith, J.V. Mannering, and W.C. Moldenhauer (1977), initiated long-term studies evaluating several tillage systems that included no-tillage and chisel-plow methods. Their studies provided much needed information on matching tillage systems to different soil and environmental conditions (Hill et al., 1994b).

In the early 1960s, George McKibben of the Illinois Research Station at Dixon Springs tested numerous planters, tillage tools, and herbicides in no-tillage experiments. He later helped introduce the double-cropping concept of no-tillage soybeans after small grains. Harry Young Jr., an innovative farmer at Hopkinsville, Kentucky, became interested in McKibben's studies and started an experiment on his own farm in 1962. Young was one of the earliest farmers to commercially adopt no-tillage, using it on nearly 500 ha. Shirley Phillips, an Extension grain crop specialist at the University of Kentucky and a dynamic spokesperson for no-tillage, was extremely successful in persuading farmers in Kentucky to adopt this revolutionary approach to the production of grain crops. Phillips and Young co-authored a book in 1973 entitled *No-Tillage Farming*.

A variety of planters and drills were manufactured in the 1970s that made it easier to plant into heavy residues as well as into dry, hard soils. During this period, post-emergence herbicides were being developed, which reduced the weed-control risk associated with no-tillage soybean. By the late 1980s, several post-emergence herbicides were available for use on corn.

The feasibility of planting into unplowed soil and leaving residues on the surface were especially enhanced by the development of these new herbicides which provided alternatives to cultivation for controlling weeds. Equipment manufacturers and innovative farmers continue to develop equipment that facilitates the management of crop residues in cropping systems.

The successful development and adoption of conservation tillage technology was most timely for U.S. farmers. Congress passed the Food Security Act in 1985 designed to conserve our soils and ensure adequate food and fiber supplies for future generations. One of the provisions of the bill is mandatory compliance with the conservation program on highly erodible land. Conservation tillage has proved to be the most cost-effective means for controlling erosion on highly erodible land (Moldenhauer and Blevins, 1995). Since passage of the Food Security Act, new research and the adoption rate of no-tillage, ridge tillage, stubble-mulch tillage, and other forms of conservation tillage have accelerated rapidly. As we approach the twenty-first century, we have the knowledge and technology to successfully grow crops using a wide variety of conservation systems.

Conservation Tillage and Soil Erosion

A major reason for using conservation tillage in the South, Southeast, Northeast, and Midwest is to control erosion by water, to save time, fuel costs, and increase profits. Because year-round soil protection is necessary to control erosion in the more humid climate regions, it is necessary that tillage practices maintain crop residues on or near the soil surface continuously. In the Southeast, sufficient residue can be produced by double-cropping grain crops (Langdale et al., 1979) or by growing cool-season legumes such as crimson clover (*Trifolium incarnatum* L.) as reported by Langdale et al. (1987).

Conservation tillage methods, when compared to inversion tillage methods, usually reduce soil erosion losses from agricultural lands but not necessarily the volume of runoff. Differences in runoff volumes have been attributed to tillage effects on soil roughness and residue cover (Romkens et al., 1973; Mannering et al., 1975; Lindstrom and Onstad, 1984). The effects of roughness are particularly apparent with chisel-plowing. Andraski et al. (1985) showed that chisel-plowing significantly reduced runoff relative to moldboard-plow tillage.

Surface water runoff generally decreases with increased amounts of surface residues, but surface residues alone may not reduce runoff if soil conditions are not favorable for water infiltration (Unger et al., 1988). Restrictive E and B horizons of Ultisols in the southeastern USA resulted in the development by Jerrell Harden of the original strip-till with in-row subsoiling used extensively in the Gulf Coastal Plains (Langdale, 1994).

Conservation Tillage Effects on Erosivity

Soil erosion is a function of erosivity and erodibility. Erosivity is related to the physical characteristics of the rainfall, especially the energy of a storm, and runoff velocity (Figure 1), and erodibility is related to the physical features of the soil and how the soil is managed.

Any management of the land that decreases the raindrop impact and resulting soil detachment will decrease erosion. This may be accomplished by a crop canopy or by plant residues (surface mulch). Erosion can also be reduced by leaving the soil rough and/or cloddy. This principle applies to wind erosion

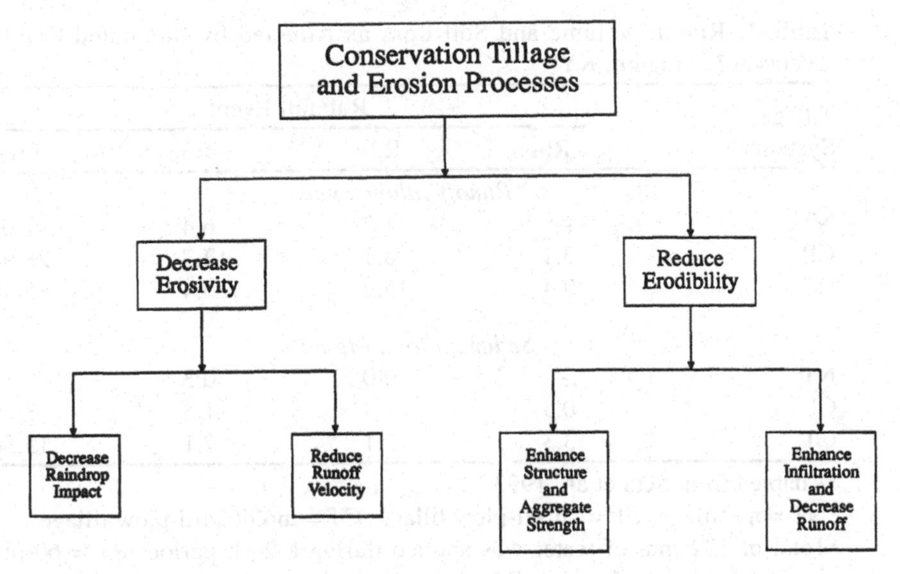

Figure 1. Mechanisms and processes involved in erosion control by conservation tillage.

as well. The many variations of conservation tillage tend to reduce soil erosion by reducing raindrop impact and runoff velocity.

One of the most dramatic examples of the influence of tillage on soil erosion is the study in Ohio reported by Harrold and Edwards (1972). On sloping watersheds planted to corn, the soil losses following a heavy rainstorm were 0.07 Mg ha^{-1} on a no-tillage watershed, 7.2 Mg ha^{-1} on a moldboard-plowed watershed planted on the contour, and 50.7 Mg ha^{-1} on a moldboard-plowed watershed not planted on the contour.

Studies conducted on sloping soils in Kentucky (Seta et al., 1993) showed that no-tillage compared to moldboard-plow tillage reduced total soil losses by 98%. Chisel-plow tillage reduced soil loss by 79% (Table 1).

Studies in Watkinsville, Georgia on highly erodible Typic Hapludults soils associated with a udic-thermic climate and highly erosive storms also make a strong case for conservation tillage (Langdale et al., 1992a). Figure 2 shows the dramatic decrease in annual soil loss from about 25 MT ha^{-1} under monocrop moldboard-plow tillage before 1975 to nearly zero as soon as conservation tillage was introduced and crop residues were returned to the soil surface at an average annual rate of 9 Mg ha^{-1}.

Conservation Tillage and Erodibility of Soils

Nature, along with humankind's management of the soil over time, either degrade or enhance the soil physical properties that determine erodibility. The

Table 1. Runoff Volume and Soil Loss as Affected by Simulated Rainfall Events at Lexington, KY[a]

Tillage System[b]	Rainfall Event			
	Rl	R2	R3	Total
		Runoff volume, mm		
NT	--[c]	1.2	6.4	7.6a[d]
CP	3.1	8.1	17.7	28.9b
CT	9.4	15.2	20.4	45.0b
		Sediment loss, Mg ha^{-1}		
NT	--	0.0	0.3	0.3a
CP	0.5	1.4	1.5	3.3b
CT	3.3	5.1	7.1	15.5c

[a]Adapted from Seta et al., 1993.
[b]NT = no-tillage; CP = chisel-plow tillage; CT = moldboard-plow tillage.
[c]Total of 132 mm of water was applied during a 24-h period; R1 = 60-min. run on moderately dry soil; R2 = 30-min. run on wet soil; and R3 = 30-min. run on a very wet soil.
[d]Means in a column followed by different letters are significantly different ($P < 0.05$).

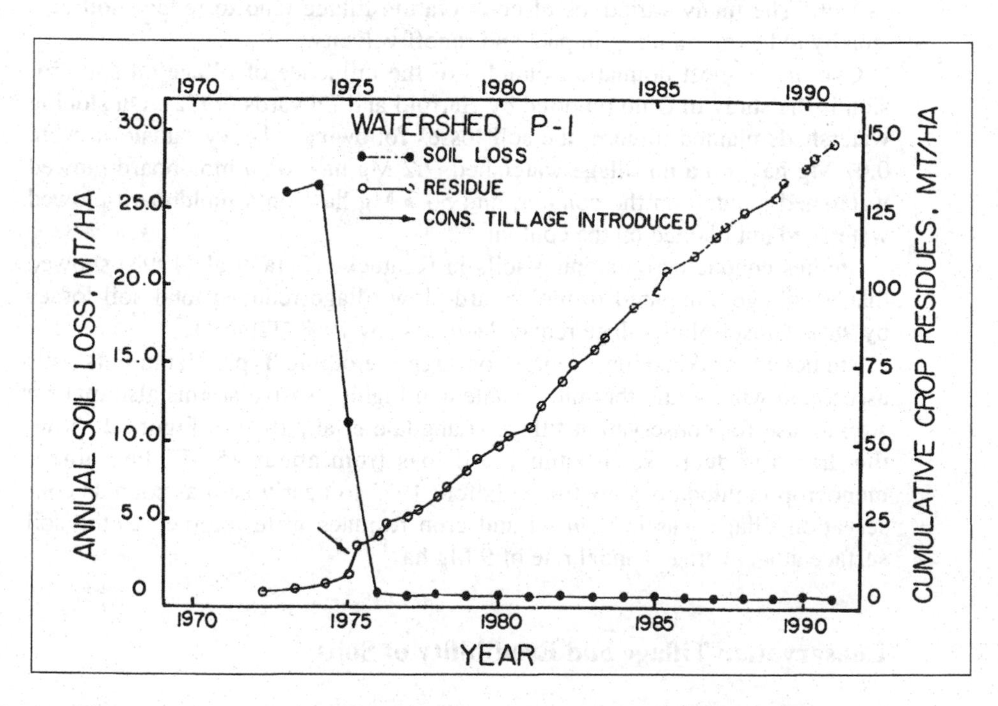

Figure 2. Effect of cumulative conservation tillage residues on annual losses. (From Langdale et al., 1992a).

organic matter content; number, size, strength, and stability of soil aggregates; and number and size of macropores are properties that influence the erodibility of a soil. Well-granulated (aggregated) soil structure that is somewhat water-stable allows movement of air and water. Soils that are at least moderately high in organic matter and have good structure tend to allow rapid entry of water into the soil. Conservation tillage that returns crop residues to the soil surface each year usually increase soil organic matter (Ismail et al., 1994; Langdale et al., 1992b) and improve soil structure.

The kind and amount of tillage determine whether soil structure is being enhanced or degraded. Along with tillage, the management of previous crop residues is a key to soil structural development and stability. If organic matter is maintained or increased and soil aggregation is enhanced, these properties directly determine the soil's capacity to infiltrate water, which in turn decreases runoff. Less runoff directly reduces soil erosion.

A long-term tillage and erosion study on soils of the Southern Piedmont in Georgia (Langdale et al., 1992b) compared organic carbon, water-stable aggregates, and rainfall retention for soil that was initially (1971–74) in conventionally tilled (moldboard-plowed) soybean to those properties after 14 years of no-tillage grain production planted into winter cover crops. Table 2 shows the dramatic changes in soil carbon from 0.7 to 2.5% along with a significant increase in water-stable aggregates with no-tillage. No-tillage also resulted in a doubling of grain yields.

Conservation Tillage for Carbon Sequestration

World soils represent a principal terrestrial carbon pool and play a major role in the global carbon cycling (Lal et al., 1995). Conservation tillage affects carbon cycling by influencing soil organic carbon content and its distribution in the soil profile, soil aggregation, and rate of soil organic carbon turnover by soil biota. Conservation tillage accentuates the rate of soil aggrading

Table 2. Effect of Tillage on Soil Carbon, Aggregate Stability, Rainfall Retention, and Grain Yield in Georgia

Carbon,[b] %	Water Stable Aggregates, %	Rainfall Retention, %	Relative Grain Yield, %
	Conventional tillage		
0.70	61	60	50
	Conservation tillage		
2.50	91	80	100

[a]Adapted from Langdale et al., 1992a.
[b]Soil depth = 0.0 to 0.15 m.

processes that enhance soil organic carbon content including: (i) humification of crop residues and other biomass, (ii) increase in nonlabile or passive soil organic carbon fraction, (iii) formation of organo-mineral complexes, and (iv) stratification of organic carbon in subsoil horizons through biochannels. A successful application of conservation tillage, especially when used on a long-term basis, also dampens the rate of soil degradative processes that diminish organic carbon content including soil erosion, surface runoff, and mineralization of organic carbon. Consequently, soil organic carbon content is usually more in soils managed with conservation than with conventional tillage (Lal et al., 1989; Dalal, 1989, 1992; Carter, 1993).

The global increase in atmospheric carbon content is estimated at 3.2 Pg (peta gram 10^{15}) C yr^{-1} (Lal et al., 1995). If conservation tillage could be adopted on 100% of the 1.5 billion ha of cultivated land area leading to 0.01% yr^{-1} increase in soil organic carbon content of the top 1 m depth with an average bulk density of 1.5 Mg/m^3, it would lead to carbon sequestration at the rate of 3.0 Pg C yr^{-1}. This change over several decades could have the potential to effectively nullify the greenhouse effect. Using the same logic, Kern and Johnson (1991) evaluated the potential impact of conservation tillage on carbon sequestration in soils of the contiguous USA. They estimated that maintaining conventional tillage level of 1990 until 2020 would result in 219 Tg (tera grams, 10^{12}) of soil organic carbon loss. In contrast, conversion of conventional to conservation tillage would result in 81 Tg of soil organic carbon loss in soil for the low scenario and 212 Tg soil organic carbon gain for the high scenario (Table 3). Therefore, widespread adoption of conservation tillage has potentially higher and more stable yields, higher profit, improved water quality, and possible mitigation of the greenhouse effect.

Table 3. Potential Impact of Conservation Tillage in Carbon Sequestration in USA[a]

	Net Carbon Gain or Loss (Tg C)		
Scenario	Mean	Minimum	Maximum
1. 1990 tillage practices continues until 2020.	−219	−203	−235
2. Conservation tillage increased to 57% of the cropland area by 2010 and then continued at that level till 2020.	−81	−96	−65
3. Conservation tillage incresed to 76% of the cropland area by 2010 and then continued at that level till 2020.	+212	+126	+298

[a]Modified from Kern and Johnson, 1991.

Conservation Tillage and Soil Quality

Numerous efforts have been made recently to define soil quality. It is defined largely by soil functions and represents a composite of its physical, chemical, and biological properties (Figure 3). Doran and Parkin (1994) defined soil quality as "the capacity of a soil to function within ecosystem boundaries to sustain biological productivity, maintain environmental quality, and promote plant and animal health."

Land management systems are sustainable only when they maintain or improve resource quality (Doran and Parkin, 1994). Soils have different levels of quality that are defined by stable natural or inherent features related to soil-forming factors and dynamic changes brought about by soil management (Pierce and Larson, 1993). The dynamic changes resulting from soil management is where conservation tillage has a direct influence on soil quality. Identifying soil properties that serve as indicators of soil quality is complicated by the large number of physical, chemical, and biological factors involved and their interaction in time and space. Thus, to evaluate the diverse effects of climate and management on soil functions requires the integration of the numerous physical, chemical, and biological indicators.

Improvements in soil attributes resulting from conservation tillage may occur because of changes in physical properties with time. For example, soil

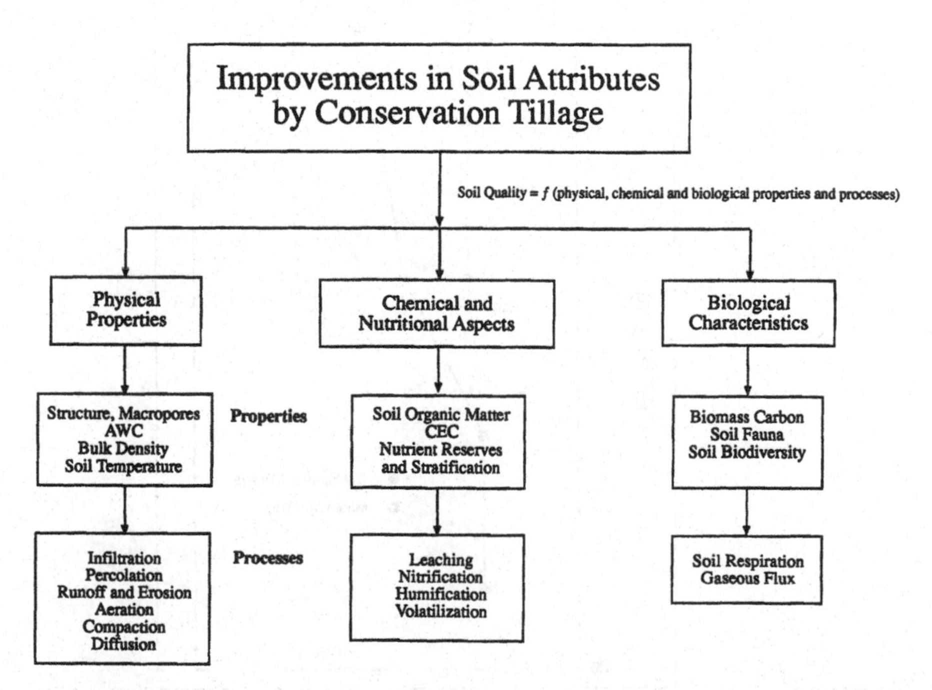

Figure 3. Conservation tillage effects on properties and processes that affect soil quality.

structure, amount and distribution of macropores, water-supplying capacity, bulk density, and soil temperature may be significantly modified by using tillage systems that decrease soil disturbance and leave crop residues on or near the soil surface.

One of the long-term effects of reduced tillage and crop residue management is an increase in soil organic matter (Edwards et al., 1992; Ismail et al., 1994; Bruce et al., 1995). Researchers in Alabama (Edwards et al., 1992) demonstrated dramatic differences in soil organic matter between conventional tillage and no-tillage in the top 10 cm of soil after 10 years of corn-wheat-soybean-wheat rotation (Figure 4). Researchers in Kentucky (Ismail et al., 1994) reported greater organic nitrogen and carbon in the 0- to 5-cm depth in no-tillage than in moldboard-plowed treatments after 20 years of continuous corn. Measurements made after 25 years on those plots show increased organic carbon levels below the 30-cm plow zone with time (unpublished data). This suggests the tillage treatments have not reached a steady state and that long-term no-tillage management is increasing the organic carbon level below the 30-cm soil depth.

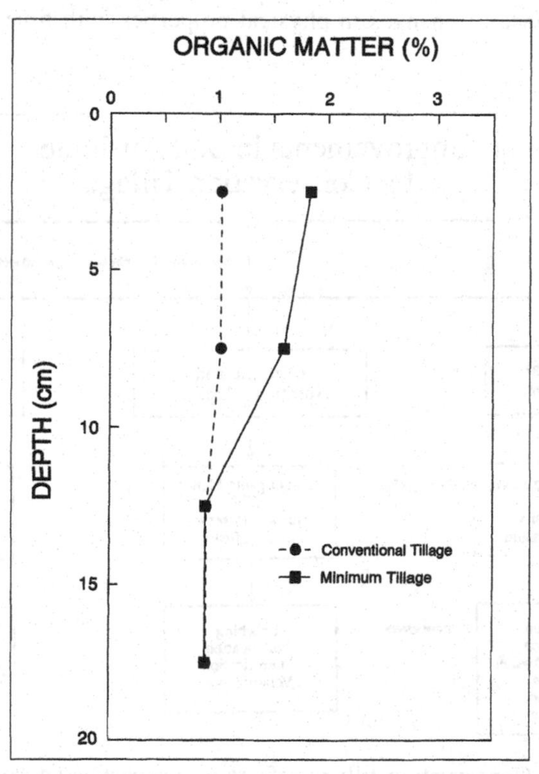

Figure 4. Residual organic matter after 10 years of corn-wheat-soybean-wheat rotation at Crossville, AL. (Adapted from Edwards et al., 1992).

Productivity and soil quality are strongly influenced by soil organic matter content. No-tillage production systems tend to reduce the rate at which organic matter is broken down by microorganisms and fauna, leaving more residue on the soil surface (Reicoskey et al., 1995).

The high-residue organic matter that accumulates in the few surface centimeters of soil under conservation tillage reduces slaking and disintegration of aggregates when they are wetted. Long-term studies in Indiana show that soil structure is better when reduced tillage systems are used that leave residues on the soil surface (Table 3). This study on four Indiana soils demonstrates that as tillage intensity is increased, soil aggregation decreases. Aggregation was highest in the 0- to 5-cm layer of no-tillage treated soil. Tillage studies on a silty soil in Germany (Ehlers, 1985) showed similar results. Research on poorly drained soils in northern Ohio (Lal et al., 1989) showed that median aggregate size tended to be considerably higher for no-tillage treatments than for plow-tillage treatments.

Karlen et al. (1994) evaluated long-term tillage studies with continuous corn and determined that, after 12 years, no-tillage resulted in soils with more stable aggregates, higher total carbon, more microbial activity, and greater earthworm populations than either chisel-plow or moldboard-plow treatments. They concluded that the higher soil quality index given to no-tillage indicated that long-term no-tillage management had improved soil quality (Table 4).

Conservation Tillage and Biological Activity

Leaving crop residues on the soil surface also increases biological activity, which is vital in improving soil quality. Soil fauna consist of a diverse group of organisms ranging in size from a few micrometers to over a meter in length (Swift et al., 1979). The activities of soil fauna directly influence certain soil processes that reflect soil quality. These processes are commonly involved in nutrient cycling and soil structural development.

Table 4. Effect of Tillage on Soil Aggregation of Four Indiana Soils After Five Years[a]

	Aggregation Index	
Tillage System	0–5 cm	5–15 cm
Conventional	0.35	0.47
Chisel plow	0.46	0.56
Till-plant	0.47	0.56
No-tillage	0.77	0.70

[a]From Mannering et al., 1975.

Biological processes associated with soil fauna include bioaccumulation of heavy metals and organic pollutants, decomposition of organic matter, mineralization of C and nutrients, and soil structure modifications from burrowing, fecal deposition, soil aggregation, and mixing of organic matter (Linden et al., 1994).

Earthworm presence and activity are more commonly identified with the above processes and usually are a good indicator of soil quality. Most farmers desire to promote the activity of earthworms and recognize their functions as beneficial (Edwards, 1989; Hendrix et al., 1992). Earthworm activity results in increased infiltration and better aeration status of a soil (Linden et al., 1994). Earthworms ingest and then transport and excrete both mineral and organic matter to other parts of the soil profile. The fecal deposits from earthworms provide an enriched environment that favors microbial activity. A major contribution of earthworms to soil quality is the conversion of plant residues into organic matter (Lee, 1985).

Selecting the proper indicators to estimate the microbial component of soil quality is somewhat difficult. Changes in soil physical and chemical properties resulting from tillage or lack of tillage can greatly alter the soil environment that supports the growth and activity of microbial populations (Doran and Linn, 1994). Microbial biomass measurements are commonly used to estimate the biological status of a soil. Anderson and Domsch (1990) suggested the use of CO_2 as an indicator of the status of soil microbial biomass and, therefore, soil quality. Turco et al. (1994) concluded in their review article on microbial indicators of soil quality that we have a variety of methods avail-

Table 5. Tillage Effects on Soil Quality in the Surface 50 mm Following 12 Years of Continuous No-Till Corn Production in Wisconsin[a]

Tillage Treatment	Wet Aggregate Stability, %	Total C in Aggregates g kg^{-1}	Biomass[b]	Respiration[c]	Earthworms[d]
			mg C kg^{-1} soil		
No-till	45.9	24	696	352	78
Chisel-plow	33.9	16	394	139	52
Moldboard-plow	35.9	11	260	74	53
LSD (0.05)	NS	4	276	114	18

[a]Adapted from Karlen et al., 1994.
[b]Biomass = microbial CO_2 evolved from soil fumigated with chloroform and reinoculated.
[c]Respiration = CO_2 evolved from untreated soil.
[d]Number of earthworms coming to the surface within a 0.25-m^2 frame after saturating soil with formaldehyde.

able to assess the microbial status of a soil, but we do not know how the combined microbial activity affects soil quality.

According to Kennedy and Papendick (1995), a healthy soil produces a healthy plant that is efficient in the accumulation of nutrients. A high quality soil is usually biologically active and contains a balanced population of microorganisms. Microbial activity can provide us evidence of subtle changes taking place in the soil long before it can be accurately measured by parameters such as organic matter. There is need for further research and effort to interpret microbial measurements and integrate this information in order to make valid soil quality assessments. For example, an increase in biomass or soil microbial activity cannot always be interpreted as an improvement in soil quality. Certainly, management practices that include changes in tillage method, crops grown, and management of crop residues all influence the soil environment and the biological processes of the soil. These practices and processes ultimately influence soil quality.

Summary and Conclusions

Conservation tillage is enjoying widespread popularity and acceptance, and this growth is expected to continue during the next decade. The progress made in the past 50 years to improve conservation tillage technology is a giant step toward our goals of conserving and wisely using soil and water resources. We now have the technology to successfully control weeds with herbicides and the equipment to plant into heavy surface residues or hard soils. With current technology, farmers have a wide range of choices in selecting a conservation tillage method that best suits their soils, climate, and cropping systems and effectively controls soil erosion.

Crop residue management systems associated with conservation tillage is the preferred method to control erosion to meet compliance requirements of the Food Security Act. Management practices that enhance soil structure formation and aggregate strength or stability will reduce soil erodibility significantly. Improved soil structure provides a soil environment that allows rapid movement of air and water into the soil. These physical changes in soil structure are generally associated with higher organic matter levels on or near the soil surface.

Conservation tillage affects global carbon cycling by influencing soil organic carbon since soils represent a principal carbon pool. If conservation tillage was adopted on all of the 1.5 billion ha of cultivated land area in the world, it would result in enough carbon sequestration to effectively nullify the greenhouse effect.

The quality of many soils in North America has declined during this century due to the use of large tillage and harvesting equipment, resulting in the degradation of soil structure, decreased tilth, water infiltration, and, in some instances, soil compaction. The same factors that enhance soil structure and decrease soil erosion are also good indicators of soil quality. The dynamic changes resulting from conservation tillage are closely related to crop residue management that directly affects the sustainability and productivity of a soil. Leaving crop residue on the soil surface protects the soil against erosion year-round and enhances biological activity and usually improves soil quality. Biological processes play a major role in decomposition of organic matter; mineralization of carbon, nitrogen, and other nutrients; and control biogeochemical cycling. A high quality soil is normally characterized as biologically active and contains a diverse population of microorganisms.

Even though research and development has made giant strides during the past fifty years, there still exists a need to modify and refine existing conservation tillage methods. We still need to control weeds more efficiently, better utilize our water resources, decrease soil erosion, and improve soil quality. Future research on conservation tillage is needed to enhance existing technology with the goal of improving, or at least maintaining, crop yields without degrading soil or water resources. We need to evaluate the impact of land use and management choices on soil processes and make adjustments as needed to avoid soil degradation and, in turn, improve soil quality. We need to better utilize available data that characterize individual components of soil quality so we can better quantify soil quality. We need to continue our efforts to better understand the relationship that exists between agriculture production method and environmental degradation. There is the need to develop a more systematic research approach to determine the soil physical, chemical, and biological properties that are influenced by conservation tillage management and determine how the interactions of these properties affect biological activity and soil quality.

Acknowledgments

Published as Paper No. (95-06-168) with approval of the Director of Kentucky Agric. Exp. Stn. Lexington.

References

Allen, R.R., and C.R. Fenster. 1986. Stubble-mulch equipment for soil and water conservation in the Great Plains. J. Soil Water Conserv. 41:11–16.

Anderson, T.H., and K.H. Domsch. 1990. Application of eco-physiological quotients (q CO_2 and q D) on microbial biomass from soils of different cropping histories. Soil Biol. Biochem. 22:251–255.

Andraski, B.J., T.C. Daniel, B. Lowery, and D.H. Mueller. 1985. Runoff results from natural and simulated rainfall for four tillage systems. Trans. ASAE 28:1219–1225.

Beale, O.W., and G.W. Langdale. 1964. The compatibility of corn and coastal bermudagrass as affected by tillage methods. J. Soil Water Conserv. 19:238–240.

Bruce, R.R., G.W. Langdale, L.T. West, and W.P. Miller. 1995. Surface soil degradation and soil productivity restoration and maintenance. Soil Sci. Soc. Am. J. 59:654–660.

Carter, M.R. (ed). 1993. Conservation tillage in temperate agro-ecosystems. Lewis Publishers, Boca Raton, FL.

Dalal, R.C. 1989. Long term effects of no-tillage, crop residue, and nitrogen application on properties of a Vertisol. Soil Sci. Soc. Am. J. 53: 1511–1515.

Dalal, R.C. 1992. Long term trends in total nitrogen of a Vertisol subjected to zero-tillage, nitrogen application, and stubble retention. Aust. J. Soil Res. 30: 223–231.

Doran, J.W. and T.B. Parkin. 1994. Defining and assessing soil quality. pp. 3-21. In: J.W. Doran et al. (ed.) Defining soil quality for a sustainable environment. SSSA Spec. Publ. 35. SSSA, Madison, WI.

Doran, J.W., and D.M. Linn. 1994. Microbial ecology of conservation management systems. pp. 1-27. In: J.L. Hatfield and B.A. Stewart (ed.) Soil biology: Effects on Soil Quality. Adv. in Soil Sci., Lewis Publishers, Boca Raton, FL.

Duley, F.L., and J.C. Russel. 1939. The use of crop residues for soil and moisture conservation. J. Am. Soc. Agron. 31:703–709.

Edwards, J.H., C.W. Wood, D.L. Thurlow, and M.E. Ruf. 1992. Tillage and crop rotation effects on fertility status of a Hapludult soil. Soil Sci. Soc. Am. J. 56:1577–1852.

Edwards, W.M. 1989. Impact of herbicides on soil ecosystems. Crit. Rev. Plant Sci. 8:221–257.

Edwards, W.M. 1995. Effects of tillage and residue management on water for crops. pp. 12–15. In: R.L. Blevins and W.C. Moldenhauer (ed.) Crop Residue Management to Reduce Erosion and Improve Soil Quality: Appalachian and Northeast. USDA-ARS, Conserv. Res. Report No. 41.

Ehlers, W. 1985. Observations on earthworm channels and infiltration on tilled and untilled loess soil. Soil Sci. 119:242–249.

Griffith, D.R., J.V. Mannering, and W.C. Moldenhauer. 1977. Conservation tillage in the eastern cornbelt. J. Soil Water Conserv. 32:20–28.

Harrold, L.L., and W.M. Edwards. 1972. A severe rainstorm test of no-till corn. J. Soil Water Conserv. 27:30.

Hendrix, P.F., B.R. Mueller, R.R. Bruce, G.W. Langdale, and R.W. Parmeke. 1992. Abundance and distribution of earthworms in relation to landscape factors on the Georgia Piedmont, USA. Soil Biochem. 24:1357–1361.

Hill, P.R., D.R. Griffith, G.C. Steinhardt, and Samuel Parsons. 1994a. The evolution and history of no-till farming in the Midwest. Natl. Conserv. Tillage Dig. 1(2):14–16.

Hill, P.R., D.R. Griffith, G.C. Steinhardt, and Samuel Parsons. 1994b. The evolution and history of no-till farming. Natl. Conserv. Tillage Dig. 1(4):14–15.

Ismail, I., R.L. Blevins, and W.W. Frye. 1994. Long-term no-tillage effects on soil properties and continuous corn yields. Soil Sci. Soc. Am. J. 58:193–198.

Karlen, D.L., N.C. Wollenhaupt, D.C. Erbach, E.C. Berry, J.B. Swan, N.S. Eash, and J.L. Jordahl. 1994. Long-term tillage effects on soil quality. Soil Tillage Res. 32:313–327.

Kennedy, A.C., and R.I. Papendick. 1995. Microbial characteristics of soil quality. J. Soil Water Conserv. 50:243–248.

Kern, J.S. and M.G. Johnson. 1991. The impact of conservation tillage use on soil and atmospheric carbon in the contiguous United States. USEPA, Corvallis, OR.

Lal, R. 1989. Conservation tillage for sustainable agriculture. Adv. Agron. 42:85:197.

Lal, R., J. Kimble, E. Levine, and B.A. Stewart (eds). 1995. Soils and Global Change. Lewis Publishers, Boca Raton, FL.

Lal, R., T.J. Logan, and N.R. Fausey. 1989. Long-term tillage and wheel-track effects on a poorly drained Mollic Ochraqualf in northwest Ohio. 1. Soil physical properties, root distribution, and grain yield of corn and soybeans. Soil & Tillage Res. 14:34–58.

Langdale, G.W. 1994. Conservation tillage development in the southeastern United States. pp. 6–9. In: P.J. Bauer and W.J. Busscher (ed.) Proceedings of 1994 Southern Conservation Tillage Conference on Sustainable Agriculture. Columbia, SC.

Langdale, G.W., A.P. Barnett, R.A. Leonard, and W.G. Fleming. 1979. Reduction of soil erosion by the no-till system in the Southern Piedmont. Trans. ASAE 22:82–86.

Langdale, G.W., M.C. Mills, and A.W. Thomas. 1992a. Use of conservation tillage to retard erosion effects of large storms. J. Soil Water Conserv. 47:257–260.

Langdale, G.W., L.T. West, R.R. Bruce, W.P. Miller, and A.W. Thomas. 1992b. Restoration of eroded soil with conservation tillage. Soil Technol. 5:81–90.

Langdale, G.W., R.R. Bruce, and A.W. Thomas. 1987. Restoration of eroded Southern Piedmont land in conservation tillage systems. pp. 142–143. In: J.F. Power (ed.) The role of legumes in conservation tillage systems. Soil Conserv. Soc. Am., Ankeny, IA.

Larson, W.E., and O.W. Beale. 1961. Using crop residues on soils of the humid area. U.S. Dept. Agric. Farmers Bull. No. 2155. U.S. Government Printing Office, Washington, D.C.

Lee, K.E. 1985. Earthworms: Their Ecology and Relationships with Soils and Land Use. Academic Press, New York.

Linden, D.R., P.F. Hendrix, D.C. Coleman, Petra C.J. van Vliet. 1994. Faunal indicators of soil quality. pp. 91–106. In: J.W. Doran, D.C. Coleman, D.F.

Bezdicek, and B.A. Stewart (ed.) Defining Soil Quality for a Sustainable Environment. SSSA Spec. Publ. No. 35. SSSA, Madison, WI.

Lindstrom, M.J., and C.A. Onstad. 1984. Influence of tillage systems on soil physical parameters and infiltration after planting. J. Soil Water Conserv. 39:149–152.

Mannering, J.V., Griffith, D.R., and Richey, C.B. 1975. Tillage for moisture conservation. Paper No. 75-2523. Am. Soc. Agric. Eng., St. Joseph, MI.

McAdams, W.N., and O.W. Beale. 1959. Wheel-track planting in mulch and minimum tillage operations. Abstract, Proc. Assoc. So. Agr. Workers.

McAllister, J.T. 1962. Mulch tillage in the southeast—planting and cultivating in crop residue. U.S. Dept. of Agric. Leaflet No. 512.

Middleton, H.E. 1952. Modifying the physical properties of soil. pp. 24–41. In: B.T. Shaw (ed.) Soil Physical Conditions and Plant Growth. Academic Press, New York.

Moldenhauer, W.C. and R.L. Blevins. 1995. Introduction. In: R.L. Blevins and W.C. Moldenhauer (ed.) Crop Residue Management to Reduce Erosion and Improve Soil Quality. USDA-ARS, Conserv. Res. Report No. 41.

Moody, J.E., G.M. Shear, and J.N. Jones, Jr. 1961. Growing corn without tillage. Soil Sci. Soc. Am. Proc. 25:516–517.

Moschler, W.W., G.D. Jones, and G.M. Shear. 1969. Stand and early growth of orchardgrass and red clover seeded after no-tillage corn. Agron. J. 61:475–476.

Nutt, G.B., W.N. McAdams, and T.C. Peele. 1943. Adapting farm machinery to mulch culture. Agric. Engr. 24:304–305.

Peele, T.C. 1942. Influences of mulches on runoff, erosion, and crop yields. S.C. Agr. Exp. Sta. Report. 1942.

Pierce, F.J., and W.E. Larson. 1993. Developing criteria to evaluate sustainable land management. pp. 7–14. In: J.M. Kimble (ed.) Proc. of the 8th Int. Soil Management Workshop: Utilization of Soil Survey Information for Sustainable Land Use. May 1993. USDA-SCS, National Soil Survey, Lincoln, NE.

Reicosky, D.C., W.D. Kemper, G.W. Langdale, C.L. Douglas Jr., and P.E. Rasmussen. 1995. Soil organic matter changes resulting from tillage and biomass production. J. Soil Water Conserv. 50:253–361.

Romkens, M.J.M., D.W. Nelson, and J.V. Mannering. 1973. Nitrogen and phosphorus composition of surface runoff as affected by tillage method. J. Environ. Qual. 2: 292–295.

Seta, A.K., R.L. Blevins, W.W. Frye, and B.J. Barfield. 1993. Reducing soil erosion and agricultural chemical losses with conservation tillage. J. Environ. Qual. 22:661–665.

Swift, M.J., O.W. Heal, and J.M. Anderson. 1979. Decomposition in Terrestrial Ecosystems. Blackwell Scientific Publication, Oxford, England.

Triplett, G.B. Jr., D.M. Van Doren Jr. and W.H. Johnson. 1964. Non-plowed, strip tilled corn culture. Trans. Am. Soc. Agric. Eng. 7:105–107.

Turco, R.F., A.C. Kennedy, and M.D. Jawson. 1994. Microbial indicators of soil quality. pp. 73–90. In: J.W. Doran, D.C. Coleman, D.F. Bezdicek, and B.A. Stewart (ed.) Defining Soil Quality for a Sustainable Environment. SSSA Spec. Publ. No. 35. SSSA, Madison, WI.

Unger, P.W., G.W. Langdale, and R.I. Papendick. 1988. Role of crop residues—improving water conservation and use. pp. 69–100. In: W.L. Hargrove (ed.) Cropping Strategies for Efficient Use of Water and Nitrogen. ASA, CSSA, SSSA Spec. Publ. No. 51. ASA, CSSA, and SSSA, Madison, WI.

Wiese, A.F., E. Burnett, and J.E. Box, Jr. 1967. Chemical fallow in dryland cropping sequences. Agron. J. 59:175–177.

Soil Management Research for Water Conservation and Quality

Paul W. Unger, Andrew N. Sharpley, Jean L. Steiner, Robert I. Papendick, and William M. Edwards

Introduction

Remains of ancient aqueducts and other water-diversion systems show that early civilizations recognized the need for adequate water to obtain favorable crop yields. Even in our country, Native Americans often diverted water from streams to their crops.

Early immigrants to North America settled mainly near the eastern coast where precipitation generally was favorable for crop production. However, farther west, plant water stress occurred frequently and droughts that lasted several years occurred occasionally. A major drought in the U.S. Great Plains and the adjacent Canadian provinces in the 1930s led to widespread soil erosion by wind. To control the erosion, stubble mulch tillage was developed, which retained crop residues on the soil surface. The residues also aided soil water conservation, which improved crop yields. Subsequently, extensive research has been conducted to further improve water conservation, which is highly important for crop production. Also, agriculture must use available water resources efficiently because of increasing competition for water from other sectors of society (municipal, industrial, recreational). Efficient and responsible water use by all sectors of society is needed to help assure availability of adequate water in the future for all users.

Whereas much water conservation research has been conducted since the 1930s and 1940s, strong emphasis on water quality began more recently. When land is farmed, chemicals such as fertilizers and pesticides often are applied to obtain optimum crop yields. While tillage helps conserve water and maximize benefits of chemical inputs, it was clear in early farming systems that more tillage and chemical inputs increased the potential for degradation of water quality. Recent public concern has led to increased agricultural research regarding the effect of land management on water quality.

Ground and surface water quality is important because leachate and runoff water from agricultural land often is the primary source of water for municipal, industrial, and recreational users. High-quality water is extremely important for many purposes, including drinking, food preparation, and industrial food processing. Good-quality water is important for power generation, fishing industries, and recreational use.

Besides soil and pesticide losses from agricultural lands, many water quality concerns center on nonpoint transport of nitrogen (N) and phosphorus (P). Due to their differing mobility in soil, N concerns revolve around nitrate leaching to groundwater, whereas P concerns focus on P transport in surface runoff.

Nitrate in water has been linked to methemoglobinemia in infants, toxicities in livestock, and water eutrophication (Amdur et al., 1991). If reduced to nitrite, it can cause methemoglobinemia that causes abortions in cattle. Phosphorus in water is not considered directly toxic to humans and animals (Amdur et al., 1991). However, free air-water exchange of N and fixation of atmospheric N by blue-green algae means that P most often limits freshwater eutrophication (Sharpley et al., 1994).

It is impractical to discuss in detail the vast literature regarding water conservation and quality. Hence, we will identify practices affecting water conservation and quality and indicate the principles involved, but give only selected examples of results that can be expected from using the different practices.

Function of Soil Management Techniques for Conserving Water

Overall goals of soil management regarding water conservation are to promote water entry into soil, reduce evaporation, and use the water to grow crops. Sometimes, excess water must be removed for successful crop production. These goals can be achieved by using appropriate tillage systems, structural and support practices, surface mulch, and cropping systems and rotations. Stewart et al. (1975) showed the relative effectiveness of various practices for reducing runoff (Table 1). Ranges in reduction given in Table 1 are shown in Figure 1.

Tillage Systems

Many tillage systems are available. We grouped them into clean, conservation, and deep tillage types to discuss effects on water conservation. Tillage influences water conservation through its effects on soil conditions that retard runoff, enhance infiltration, suppress evaporation, and control weeds. Runoff is retarded and infiltration is enhanced when water flow into soil is unrestricted by surface conditions, water is temporarily stored on the surface to provide more time for infiltration, and water movement within the soil profile is not impeded. Evaporation is suppressed by insulating and cooling the soil surface, reflecting solar energy, decreasing wind speed at or near the soil surface, and providing a barrier against water vapor movement. Timely weed control is highly important because weeds may deplete soil water supplies.

Table 1. Practices for Controlling Direct Runoff and Their Highlights[a]

Runoff Control Practice	Effect on Runoff[b]
No-tillage planting in prior crop residues	Variable effect on direct runoff — from substantial reductions to increases on soils subject to compaction
Conservation tillage	Slight to substantial reduction
Sod-based rotations	Substantial reduction in sod year; slight to moderate reduction in row-crop year
Meadowless rotations	None to slight reduction
Winter cover crop	Slight increase to moderate reduction
Improved soil fertility	Slight to substantial reduction, depending on existing fertility level
Timing of field operations	Slight reduction
Plow-plant systems	Moderate reduction
Contouring	Slight to moderate reduction
Graded rows	Slight to moderate reduction
Contour strip cropping	Moderate to substantial reduction
Terraces	Slight increase to substantial reduction
Grassed outlets	Slight reduction
Ridge planting	Slight to substantial reduction
Contour listing	Moderate to substantial reduction
Change in land use	Moderate to substantial reduction
Other practices	
Contour furrows	Moderate to substantial reduction
Diversions	No reduction
Drainage	Increase to substantial decrease of surface runoff
Landforming	Increase to slight decrease
Construction of ponds	None to substantial reduction

[a]From Stewart et al., 1975.
[b]Ranges in percent reduction of potential direct growing season runoff for the descriptive terms, "slight," "moderate," and "substantial," are shown in Figure 1.

Where excess water must be removed, installation of drainage systems may be necessary.

Clean tillage is the process of plowing and cultivating to incorporate crop residues and control weeds (SSSA, 1987). Water conservation with clean tillage results primarily from disrupting soil crusts, providing for temporary water storage, and controlling weeds. Under some conditions, clean tillage also suppresses evaporation.

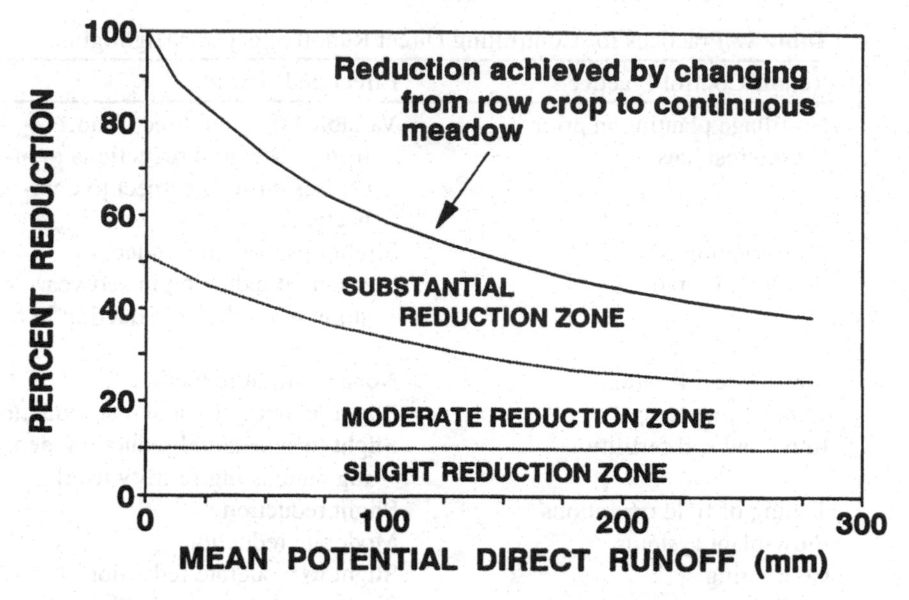

Figure 1. Ranges in percent reduction of potential direct mean growing season runoff resulting from practices shown in Table 1. (Adapted from Stewart et al., 1975).

When raindrops strike a bare soil, a surface seal often develops, resulting in reduced infiltration. When the soil dries, a crust develops that can hinder infiltration of the next rain. The residue-free surface condition produced by clean tillage often aggravates the crusting problem.

Tillage-induced surface roughness and cloddishness can reduce runoff velocity and create depressions for temporary water storage, thereby providing more time for infiltration. Tillage-induced soil loosening can increase water storage in the tillage layer (Burwell et al., 1966).

After wetting a bare soil, evaporation initially occurs at the potential rate, then becomes slower, depending on the rate of soil water movement to the surface. Disrupting water movement to the surface is one way to reduce evaporation. Shallow tillage for creating a "dust" mulch to reduce evaporation generally is ineffective where precipitation occurs mainly during the summer when the potential for evaporation is greatest and tillage is needed after each rain to control weeds (Jacks et al., 1955). Such mulch, however, reduces evaporation where a distinct dry season follows a wet period that has recharged the soil profile with water (Hammel et al., 1981; Papendick et al., 1973).

Weeds compete with crops for water, nutrients, and light, with competition for water generally being most important under dryland conditions. Therefore, effective weed control is essential if crops are to produce at their potential under the prevailing conditions.

Conservation tillage is any tillage sequence that provides at least 30% cover of crop residues on the soil surface after crop planting to control water erosion. Crop residues equivalent to at least 1.1 Mg ha^{-1} of straw must be present during the major wind erosion period to control wind erosion (CTIC, 1990). This definition emphasizes crop residue management, which is the term being used in some cases (Stewart and Moldenhauer, 1994). However, we will use conservation tillage, examples of which are stubble-mulch tillage, reduced tillage, and no-tillage.

Stubble-mulch tillage (SMT) was developed to combat wind erosion in the U.S. Great Plains and Canada in the 1930s. With SMT, sweeps or blades undercut the soil surface to sever weed roots and prepare a seedbed. Because SMT does not invert soil, most crop residues remain on the surface to enhance erosion control and provide water conservation benefits. Based on crop yields, SMT is better adapted to drier than to more humid regions, possibly because of an improved crop water-nutrient balance (Zingg and Whitfield, 1957) and better weed control in drier regions.

Since controlling weeds is a major reason for tillage, then the need for tillage is reduced if weeds are controlled by other means, as with herbicides. Reduced tillage systems that usually meet the requirements for conservation tillage include fall (autumn) chisel-field cultivate, disk-plant, till-plant, strip tillage, and tillage-herbicide combinations. Tillage-herbicide systems have received much attention where residue production is low, erosion potential is high, water conservation is important, and persistent weeds cannot be effectively controlled by tillage or herbicides alone. These systems have improved erosion control, water conservation, and crop yields, especially where precipitation is limited (Papendick and Miller, 1977; Smika and Wicks, 1968; Unger, 1984).

With no-tillage (NT), crops are planted with no preparatory tillage since harvest of the previous crop. Herbicides are used to control weeds. While NT is widely promoted to control erosion because it retains nearly all crop residues on the surface, it also provides water conservation benefits. Surface residues dissipate energy of falling raindrops, thus reducing aggregate dispersion and surface sealing and maintaining favorable water infiltration rates (Bruce et al., 1995; Unger, 1992); retard the rate of water flow across the surface, thus providing more time for infiltration; promote biological and fauna activity in soils, thus improving soil conditions for more rapid water infiltration and distribution within the soil profile (Edwards et al., 1988a, b); and reduce evaporation (Steiner, 1989). No-tillage generally is well-suited for use on well-drained and moderately well-drained soils, provided adequate residues are available and the soil is not severely degraded (Charreau, 1977). Results with NT often are poor on poorly-drained soils (Triplett and Van Doren, 1977) because reduced runoff and evaporation aggravate the poorly-drained condition.

Ridge tillage (RT) for which the seedbed level is raised above that of the surrounding soil, has become popular for producing some row crops. Use of RT aids soil drainage, improves residue management, provides for residue cycling in the plant root system, and generally is better than NT on poorly-drained soils. Other advantages include earlier soil warming, good erosion control, more timely planting because intensive tillage is not needed, and potentially less soil compaction because traffic can be confined to certain furrows. All these factors can improve soil water conservation.

Deep tillage generally means plowing 0.40 to 0.90 m deep (Burnett and Hauser, 1967). Profile modification to even greater depths is done with special equipment (Eck and Taylor, 1969). These operations improve water conservation primarily by disrupting naturally-dense or compacted soil layers that impede water movement, thus improving infiltration and increasing the depth to which plant-available water can be stored; and from mixing soil layers (for example, clayey and sandy layers), thus increasing the soil's water-holding capacity.

Structural and Support Practices

Use of appropriate tillage methods can improve water conservation on lands that have few limitations. However, as severity of limitations increases, tillage alone may become ineffective, and structural and support practices that complement tillage may be needed for effective water conservation.

For many U.S. locations, use of contour tillage reduced annual runoff by up to 20% and growing-season runoff by up to 33% (Stewart et al., 1975). Although used primarily for water erosion control, contour tillage helps conserve water because the contour ridges hold water on the entire field and provide more time for infiltration. Whereas contour tillage helps hold water on the land, graded furrows help remove water from land at nonerosive rates. Water conservation benefits may result when furrow gradients are low because runoff is slow and the potential runoff water is more uniformly distributed over the entire field (Richardson, 1973).

With basin listing, small earthen dams in furrows hold precipitation where it falls, which often prevents runoff and provides more time for infiltration (Jones and Clark, 1987). In some years, basin listing improved water storage and crop yields. Lack of response in other years resulted from inadequate rain to cause runoff, water loss by evaporation, and abundant rainfall that provided adequate water, even with unblocked furrows. Although used primarily to conserve water for dryland crops, basin listing also is an integral component of the low energy precision application (LEPA) irrigation system developed by Lyle and Bordovsky (1981). Irrigation application efficiencies above 95% have been achieved with the LEPA system.

Although used mainly to control erosion, strip cropping improves water conservation by causing water to flow through the strip of protective crops at a reduced rate, thus causing sediments to settle from the water and providing more time for infiltration. Water conservation due to strip cropping per se generally is variable, with wind speed, soil type, strip (barrier) type, climate, and crops grown affecting the results (Black and Aase, 1988; Rosenberg, 1966). In contrast, strip cropping conserves water from windblown snow by trapping snow in crop residues (often small grain stubble) or in specially-planted barrier strips (Black and Siddoway, 1977; Staple et al., 1960).

Terraces may have level or graded channels. Level terraces retain water on land whereas graded terraces remove excess water from land at a nonerosive velocity, but graded terraces have some water conservation benefits. Level terraces often have blocked outlets to prevent runoff. When blocked, they should be large enough to retain all storm water on land until it infiltrates. A disadvantage of level terraces, especially those with blocked ends, is that ponded water in channels may delay field operations or damage crops, unless the water is drained (Harper, 1941). Conservation bench terraces (CBTs) are special level terraces for which the adjacent upslope portion of the terrace interval is leveled. Runoff from the remaining nonleveled part of the terrace interval is captured and spread over the leveled area (Zingg and Hauser, 1959). A major constraint to using CBTs is the high cost of terrace construction and land leveling. To reduce these costs, Jones (1981) developed narrow CBTs (about 10 m wide), which provided water conservation benefits similar to those with larger systems (about 60 m wide).

Although designed primarily to control erosion by conveying water from land at nonerosive flow rates, use of graded terraces improves water infiltration because of the low flow rates within the terrace channel. However, most water is stored in or near the terrace channel and is of limited value for the crop on much of the field. Greater water conservation is achieved by using graded terraces along with graded furrows (discussed earlier) because increased water storage occurs on a larger part of the field (Richardson, 1973).

In contrast to CBTs for which only part of the land is leveled, the entire interval is leveled for a level bench terrace system. This system is widely used in some countries on steep slopes where cropland is limited and precipitation is relatively high, but generally is not practical for mechanized farming. Jones (1981), however, developed a level bench terrace system for conserving water for dryland crops on gently-sloping land (about 1%). By using a 5-m terrace interval, only a small amount of soil had to be moved, which greatly reduced the land-leveling cost.

Mulch

Water conservation by using mulch results from reduced evaporation, surface protection that results in more favorable infiltration, and reduced runoff rates that provide more time for infiltration. Various materials are used as mulch, but crop residues are used most commonly. Crop-residue mulch characteristics affecting evaporation include residue amount, orientation, uniformity, rainfall interception, reflectivity, and dynamic roughness (Van Doren and Allmaras, 1978).

Mulch effects on evaporation are readily apparent under laboratory conditions, but long-term effects of mulch on evaporation are difficult to show, especially under field conditions, because of its interacting effect on water infiltration, distribution in the profile, deep percolation, and subsequent evaporation.

Soil Surface Amendments

Surface sealing due to raindrop impact or flowing water on bare soils having low-stability surface aggregates can result in major losses of water as runoff. Therefore, if aggregates could be made more stable, runoff could be reduced, which could improve water conservation. By applying phosphogypsum at 10 Mg ha^{-1} to a ridged sandy soil, runoff was sixfold less than where it was not applied (Agassi et al., 1989).

Aggregate breakdown in furrows results in low infiltration and high sediment losses during furrow irrigations of crops under some conditions. When a polyacrylamide or starch copolymer solution was injected into irrigation water at different rates, net infiltration increased 15% and sediment loss decreased 94% (Lentz and Sojka, 1994). Injection of polyacrylamide at 10 g m^{-3} was one of the most effective treatments. The injection also improved lateral infiltration, which can improve water and nutrient use efficiency by row crops (Lentz et al., 1992).

Cropping Systems and Rotations

Besides effects of tillage systems, support practices, mulch, and surface amendments, water conservation is affected by the overall crop management systems in which the above practices are used. Crop management embraces such topics as management of planting materials, land use before planting, seedbed preparation, planting, plant pests, and plant products (Sprague, 1979). Subtopics related to several of the above are cropping systems and rotations, which often have a major effect on water conservation. Crop selection for a given locale generally is based on the probability of precipitation and amount of stored soil water being available to produce a satisfactory yield (Stewart and Steiner, 1990), but crops grown also influence water storage. In general, water storage should be greater for large-seeded crops that can be planted in a residue-covered or rough and cloddy seedbed than for those requiring a smooth, residue-free seedbed consisting of fine soil materials (small-seeded crops). The latter seedbeds often seal and crust severely when rain occurs, thus reducing infiltration and water conservation.

Crop growth habit and canopy influence water conservation through their effect on interception of raindrops, resistance to water flow across the soil surface, and evaporation. Upright-growing plants provide little surface protection early in the growing season, but may fully protect the surface when the canopy is complete. Low-growing vines or stoloniferous plants may provide relatively little surface cover, but retard water flow and, thereby, enhance water conservation. Fibrous-rooted plants generally provide greater stability to soil and, therefore, greater water conservation than tap-rooted plants. Densely planted crops generally provide ground cover more quickly than sparsely planted crops, which can improve water conservation due to greater infiltration and lower evaporation. Closely spaced plants also retard runoff, which provides more time for infiltration, than widely spaced plants.

Since soil water storage is influenced by the storage capacity, water remaining in soil from a previous storage event reduces additional storage. As rooting depth and intensity of water extraction increase, the potential for storing more water increases. Growing a deep-rooted crop after a shallow-rooted crop can improve water conservation (Stewart and Steiner, 1990).

Timing and duration of a crop's growing season relative to the time of most-probable precipitation can greatly influence soil water storage. For the dryland winter wheat-grain sorghum crop rotation for which a fallow period of 10 to 11 months precedes each crop in a 3-year period in the southern Great Plains, runoff is low after wheat harvest in June or July, although most rainfall occurs during the summer months (Jones and Hauser, 1975). Runoff is low because wheat usually extracts most available water, which results in a "dry" soil with a relatively large water storage capacity. In contrast, runoff during the same period is greater from land planted to grain sorghum because the soil contains water stored during the previous fallow period, and there is little opportunity for additional storage. Although antecedent soil water contents influenced the above results, differences in surface residue amounts and type also influenced the results (Jones et al., 1994).

For an irrigated crop, timing of the last irrigation greatly affects the soil water content at harvest. If the last irrigation is applied so that relatively little water remains at harvest, potentially more water can be stored during the ensuing interval between crops (Musick, 1970).

Legume or grass cover crops influence water conservation mainly through their effect on surface cover during their growth period, residues remaining after growth is stopped, use of soil water, and soil conditions resulting from their use. Although their use generally improves water conservation in more humid areas, their use in drier regions often is detrimental to water conservation because they use water that could be used by a subsequent crop.

Water conservation benefits from using legumes and cover crops may be realized also from improved soil conditions (Langdale et al., 1985; Stewart et al., 1975). These benefits result mainly from greater soil aggregate stability, which reduces aggregate dispersion and surface sealing due to raindrop impact.

Continuous cropping usually is practiced where precipitation recharges the soil profile with water between crops or supports a crop during its growing season, and generally refers to growing the same crop on the same land during successive growing seasons. We expand the meaning to include growing various crops on the same land in successive growing seasons, for example, summer crops such as corn (*Zea mays* L.), soybean (*Glycine max* L. Merr.), or sugar beet (*Beta vulgaris* L.), but without a fallow period. In contrast, crop-fallow systems are those for which land remains idle during all or the greater part of a typical growing season (SCSA, 1976; now SWCS).

A major reason for including a fallow period is to increase the soil water content at planting and, thereby, to reduce the risk of crop failure. Where precipitation is adequate, soil management usually has relatively little effect on soil profile water content at planting. As a result, a fallow period is seldom used in humid or subhumid regions. In contrast, systems involving fallow often are used in drier regions, which increases the potential for achieving favorable yields (Black, 1985).

Soil water storage generally increases as fallow period length increases. However, fallow storage efficiency, namely, unit of water stored per unit of precipitation during the fallow period, generally decreases as length of fallow increases. These trends occur because greater storage usually occurs when the antecedent soil water content is low early in the fallow period. Later, the antecedent water content is greater, which makes storing additional water more difficult.

Function of Soil Management Techniques for Protecting Water Quality

Generally, as the degree or intensity of agricultural management increases, loss of N through leaching to groundwater and P in surface runoff increases (OECD, 1982; Sharpley et al., 1994). As a result, more emphasis recently has been placed on reducing N and P losses via improved soil management. This has focused on managing soil water, residues, and cropping systems to enhance water- and nutrient-use efficiency. Soil management also influences the amount of sediments transported by runoff to surface waters.

Soil Water Management

Tillage influences surface hydrology and, thereby, soil-water budgets (Follett et al., 1987). As a result, tillage also affects transport of N and P in ground- and surface water, but effects of tillage on nitrate leaching are variable. Nitrate loss from Maury silt loam (Typic Paleudalf) in Kentucky was greater with no-tillage than with conventional tillage (Tyler and Thomas, 1977), but less nitrate leached from Clarion-Nicollet loam (Typic Hapludoll-Aquic Hapludoll) in a corn-soybean rotation in Iowa with no-tillage than with moldboard plowing (Kanwar et al., 1985).

Smith et al. (1987) and Sharpley and Smith (1993) measured nitrate concentrations in wells on watersheds from 1983 to 1994 at El Reno, Oklahoma. Annual rainfall averaged 740 mm and the water table was at depths between 3

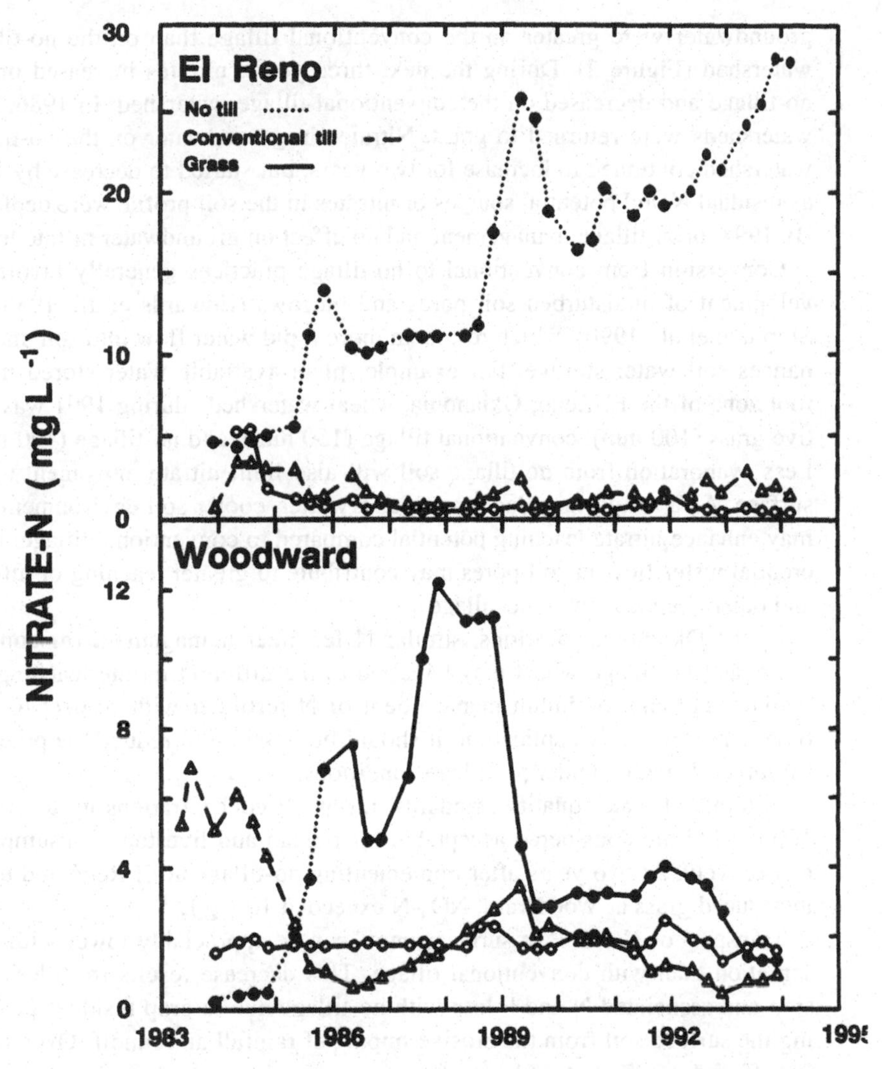

Figure 2. Nitrate-N concentrations of groundwater as a function of agricultural management at El Reno and Woodward, Oklahoma, from 1983 to 1994.

and 25 m. After converting a watershed from conventional tillage to no-tillage in 1984, nitrates gradually increased (Figure 2). Nitrate levels on the conventional and native grass watersheds were similar and consistently lower than on the no-tillage watershed.

At Woodward, Oklahoma, wells were installed in 1983 and one watershed was converted from conventional tillage wheat to no-tillage wheat the next year. Between 1983 and 1994, annual rainfall averaged 600 mm and the water table was at depths between 3 and 25 m. Early in 1984, nitrate levels in groundwater were greater on the conventional tillage than on the no-tillage watershed (Figure 2). During the next three years, nitrates increased on the no-tillage and decreased on the conventional-tillage watershed. In 1986, both watersheds were returned to grass. Nitrates in groundwater on the no-tillage watershed continued to increase for two years, but started to decrease by 1989 as residual N and potential sources of nitrates in the soil profile were depleted. By 1994, prior tillage management had no effect on groundwater nitrate level.

Conversion from conventional to no-tillage practices generally favors development of undisturbed soil pores and burrows (Edwards et al., 1988a, b; Shipitalo et al., 1990), which results in more rapid water flow into soil and enhances soil water storage. For example, plant-available water stored in the root zone of the El Reno, Oklahoma, wheat watersheds during 1991 was: native grass (100 mm), conventional tillage (150 mm), and no-tillage (190 mm). Less evaporation from no-tillage soil will also limit nitrate movement to the surface. As a result, no-tillage provides a wetter, cooler soil environment that may enhance nitrate leaching potential compared to conventional tillage. Preferential water flow in soil pores may contribute to greater leaching of nitrates and other chemicals with no-tillage.

At the Oklahoma locations, similar N fertilizer management for conventional and no-tillage wheat may have led to the different nitrate leaching potentials. Through optimum management of N fertilizer with respect to rate, time, and form of N application, it should be possible to reduce the potential for nitrate leaching under no-tillage conditions.

In terms of water quality standards, nitrate-N concentrations up to 10 and 100 mg L^{-1} are considered acceptable for human and livestock consumption, respectively. At two years after implementing no-tillage at El Reno and under introduced grass at Woodward, NO_3-N exceeded 10 mg L^{-1}.

Transport of N and P in surface runoff can be appreciably lower with conservation than with conventional tillage. This decrease results from less erosion and associated N and P loss with no-tillage due to crop residues protecting the surface soil from the erosive impact of rainfall and runoff. Over 12 yr, runoff and erosion were lower from conservation than from conventional tillage grain sorghum and wheat watersheds in the Southern Plains (Table 2). Soil-water management through use of no- and reduced-tillage practices also decreased total N and P losses in runoff (Table 2).

Although total N and P losses in runoff are lower with no-tillage than with conventional tillage, the bioavailability of P transported from no-tillage areas can be larger (Table 2). For example, bioavailable P was 82% of total P loss from wheat plots with no-tillage, but only 19% with conventional tillage at similar rates of P fertilization at El Reno, Oklahoma (Sharpley et al., 1992b), with the increased bioavailability attributed to leaching of P from crop residues and preferential transport of clay-sized particles in runoff (Andraski et al., 1985; Sharpley, 1981). Therefore, an increase in bioavailability of P transported due to using certain management practices may not reduce the

Table 2. Effect of Tillage Practice on Soil, N, and P Loss in Runoff from Sorghum and Wheat in the Southern Plains[a]

Crop and Tillage System	Runoff mm	Soil Loss kg ha^{-1} yr^{-1}	N Loss Nitrate kg ha^{-1} yr^{-1}	N Loss Total kg ha^{-1} yr^{-1}	P Loss Dissolved kg ha^{-1} yr^{-1}	P Loss Total kg ha^{-1} yr^{-1}
Sorghum						
No-	31	280	0.35	1.11	0.08	0.28
Reduced	21	520	0.41	1.40	0.04	0.37
Conventional	121	16150	0.62	1.34	0.24	4.03
Wheat						
No-	77	540	1.52	5.12	0.53	0.98
Reduced	61	800	1.92	4.59	0.10	0.59
Conventional	101	8470	1.74	20.19	0.21	3.96
Native grass	92	43	0.38	1.11	0.12	0.20

[a]Adapted from Sharpley et al., 1991; Smith et al., 1991.

trophic state of a water body as much as may be expected from inspection of total P loads only.

In terms of production, wheat grain yields were lower following implementation of no-tillage in 1983 than with conventional tillage (Sharpley and Smith, 1993). The lower yield is attributed in part to stratification of broadcast fertilizer, particularly P, in the surface 5 cm of soil. No-tillage wheat may respond to subsurface P applications and light tillage that incorporates surface-bound nutrients. Increased competition from weeds also has occurred since conversion to no-tillage. Although yields are not reduced by using no-tillage in most of the country, this Oklahoma example demonstrates the potential economic and environmental conflicts involved in soil-water management. Therefore, recommendations to farmers regarding soil-management techniques must be flexible and site-specific, addressing not only crop production goals but also the vulnerability of local water resources to either ground- or surface water impacts.

Residue Management

Crop residue management affects soil nutrient cycling and sediment transport, and potentially can influence water quality. Factors involved include quantity, type, and placement of crop residues. The quantity of residues involved will affect the amounts of nutrients being cycled and, after going through the mineralization-immobilization process, the amounts potentially available for uptake or movement with surface water. Increasing residue amounts can also reduce evaporation and keep the surface soil moist, particularly if residues are left on the surface. Thus, microbial transformation of N may increase, resulting in greater mineralization, availability, and uptake of indigenous soil N (Table 3). Greater residue P mineralization has been observed also for surface-incorporated residues (Sharpley and Smith, 1989).

The chemical composition of crop residues is particularly important in establishing the balance between mineralization and immobilization processes. Generally, if the C:N ratio is >30, net immobilization of N in residues occurs; with ratios <25, net mineralization occurs. Similarly, C:P ratios >300 favor immobilization and those <200 favor mineralization. Power et al. (1986) demonstrated that little N in corn residues (wide C:N ratio) was mineralized and used by the next crop, whereas a large part of the N in soybean residues (narrow C:N ratio) was available to the next crop (Table 3).

Different tillage implements result in crop residue placement at different depths (Staricka et al., 1991). When residues are incorporated, N and P are concentrated where the residues are placed by the given operation. In contrast, use of no-tillage concentrates nutrients near the soil surface. Due to the sorption and subsequent immobility of P in soil, P can rapidly accumulate and stratify at the surface of no-tillage soils (Griffith et al., 1977; Guertal et al., 1991).

Table 3. Uptake of N from Various Sources and Total Uptake at 1981 Harvest of Maize and Soybean (Whole Plant) as Affected by Crop Residue Rate on the Soil Surface[a]

	Source of N				
Crop and Residue Rate (%)[b]	Crop Residue	Residual Fertilizer kg ha^{-1}	Current Fertilizer kg ha^{-1}	Native Soil N[c] kg ha^{-1}	Total Uptake kg ha^{-1}
Maize					
0	0	5	4	73	82
50	0	6	7	97	110
100	2	6	7	114	129
150	1	6	11	124	142
Soybean					
0	0	2	14	84	100
50	1	2	21	124	148
100	38	7	16	116	177
150	63	6	20	106	195

[a]Adapted from Power et al., 1986.

[b]Based on the amounts of residue produced the previous year. For 100% treatment, amounts on a dry weight basis were 4.97 Mg ha^{-1} for maize and 4.58 Mg ha^{-1} for soybean.

[c]For soybean, native soil N includes biologically fixed N.

Clearly, residue management can affect N and P movement in ground- and surface water. Mostaghimi et al. (1988, 1992) found both tillage and amount of rye residues present affected N and P in runoff from a Groseclose silt loam (Typic Hapludult) in Virginia (Table 4). For both conventional and no-tillage, increasing residue levels decreased runoff and erosion. Also, N and P losses were consistently less with no-tillage than with conventional tillage. However, an increase in residues from 750 to 1500 kg ha^{-1} resulted in greater N and P losses in runoff, which were attributed to greater leaching of nitrates and dissolved P from the residues and less sorption of P by eroding soil at the higher residue level. Differences in amounts of P leached from various crop residues also affected seasonal and spatial variability in P losses among watersheds (Burwell et al., 1975; Sharpley, 1981).

Land application of materials such as manures and organic wastes can also affect ground- and surface-water quality. Factors such as composition and rate, placement, and timing of application of these materials influence N and P cycling. They also affect availability of N and P from these sources. Other organic residues important in localized areas include sewage sludge from municipalities, wastes from livestock slaughtering facilities, and wastes from food processing and other industries. If properly handled, they often serve as major N and P sources (Sharpley et al., 1995). If organic residues are added continually in amounts that provide more N and P than those removed by

Table 4. Effect of Tillage Method and Residue Amount on Soil, N, and P Loss in Runoff from Sorghum and Wheat in the Southern Plains

Tillage Method and Residue Amount (kg ha^{-1})	Runoff mm	Soil Loss kg ha^{-1} yr^{-1}	N Loss: Nitrate	N Loss: Total	P Loss: Dissolved	P Loss: Total
			kg ha^{-1} yr^{-1}		kg ha^{-1} yr^{-1}	
Conventional tillage						
0	36	2812	0.285	4.562	0.506	5.235
750	33	1001	0.283	1.665	0.265	0.982
1500	18	513	1.326	4.382	0.412	1.425
No-tillage						
0	5	72	0.210	0.608	0.073	0.101
750	3	11	0.003	0.009	0.002	0.057
1500	1	7	0.106	0.313	0.027	0.097
Reduction with no-tillage (%)						
0	87	97	26	87	86	98
750	92	99	99	99	99	94
1500	99	99	92	93	93	98

[a]Adapted from Mostaghimi et al., 1988, 1992.

crops, N and P may accumulate in soil profiles and lead to enrichment of P in runoff and nitrates in groundwater (Sharpley et al., 1995). Management of organic residues by incorporation with tillage redistributes N and P throughout the plow layer, which enhances uptake by crops. Also, timing of organic residue application to coincide with maximum crop uptake and low rainfall-runoff incidence reduces the potential for ground- and surface water contamination (Edwards et al., 1992).

Cropping System Management

Management of cropping systems can influence water quality through selection of species that maximize uptake of soil N and P, thereby reducing residual nutrients available for movement in ground- and surface water. This is most commonly accomplished through crop rotation and cover crop management.

The sequence of crops in a rotation influences available water and N movement through the soil profile and ultimately into groundwater (Carter et al., 1991). For example, legumes can effectively use or "scavenge" N remaining in soil from previous crops (Olson et al., 1970). To illustrate this, Sharpley et al. (1992a) overlaid hypothetical root development patterns for a corn-winter wheat-alfalfa (*Medicago sativa* L.) cropping system on typical N leaching patterns (Figure 3).

Olson et al. (1970) found nitrate concentrations at a depth of 1.2 to 1.5 m in a silt loam soil to be lower for an oat (*Avena sativa* L.)-meadow-alfalfa-corn rotation than for continuous corn when ammonium nitrate was applied to both systems. Nitrate reduction was directly proportional to the years oat, meadow, or alfalfa were in the rotation with corn, which was attributed to the combined recovery of nitrate by shallow-rooted oat followed by deep-rooted alfalfa.

Much information documents the benefits for N-use efficiency and reduced nitrate leaching potential with careful selection and sequencing of crops in a rotation, but less information is available for P. However, it is possible that selecting and using crops with a high affinity for P may reduce soil P stratification and increase P-use efficiency, particularly if the nonharvested part of crops is returned to the soil.

Including cover crops in management systems can affect ground- and surface water quality by reducing runoff and erosion, improving soil structure and tilth, fixing atmospheric N, and reducing nitrate leaching. Cover crops affect nitrate leaching and groundwater quality by modifying soil-water budgets and N uptake (Meisinger et al., 1991; Sharpley and Smith, 1991). Through evapotranspiration, cover crops reduce the amount of water available for leaching. However, as for crop rotations, cover crops must be carefully managed to avoid reducing soil-water reserves for the next crop. Both nonlegume and legume cover crops extract and incorporate mobile soil nitrate into immobile biomass N, thereby reducing the amount available for leaching.

Due to the above factors, inclusion of cover crops in management systems

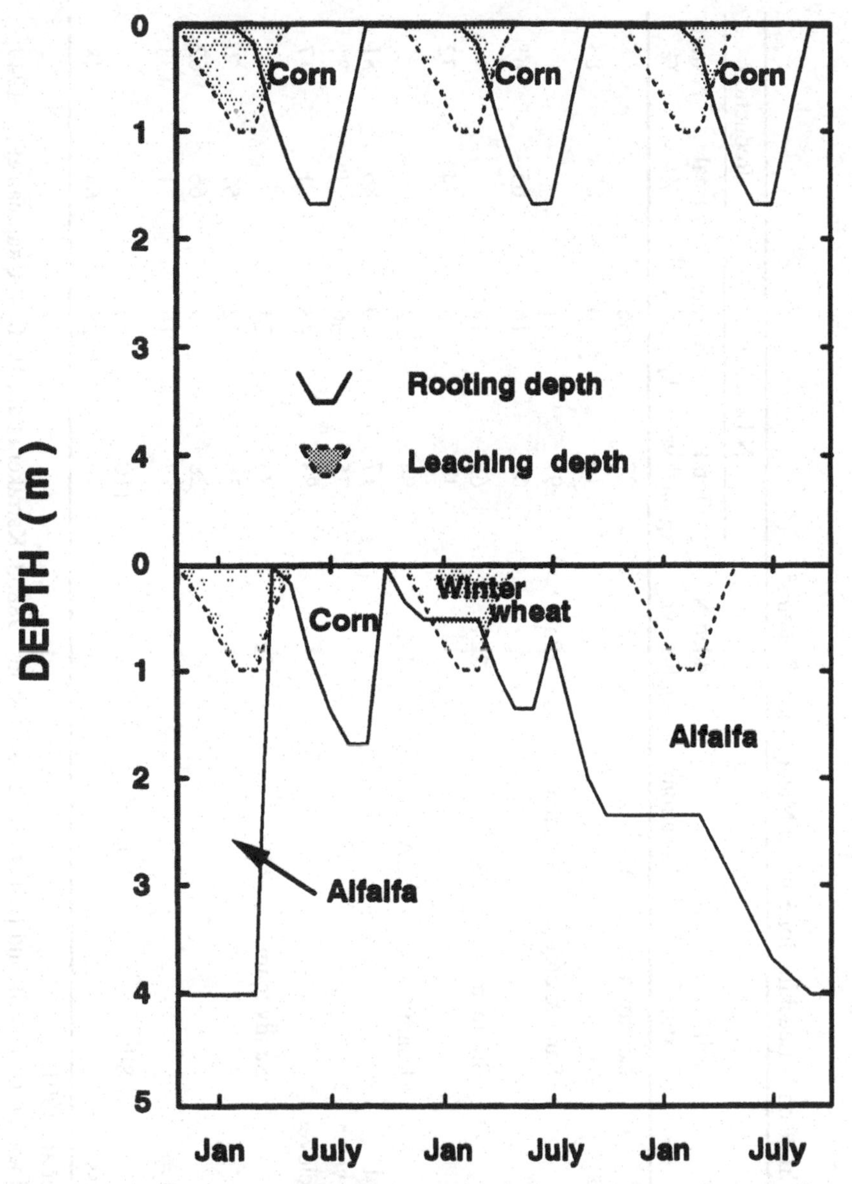

Figure 3. Typical root development of continuous corn and corn-winter wheat-alfalfa rotation in relation to soil drainage over a 3-yr period.

has consistently decreased nitrate leaching (Table 5), sometimes as much as 83%. Smaller reductions often result from winter kill and incorporation by tillage before planting the next crop and lower N uptake, if the crop is a legume.

Nonlegume cover crops can reduce soil nitrates, while legumes can provide N for the next crop. Therefore, legumes may not reduce nitrate leaching as ef-

Table 5. Effect of Cover Crops on N Leaching for Several Management Systems[a]

Crop	Cover Crop	Soil Texture	Location[b]	Added N kg ha^{-1} yr^{-1}	N Leached Total kg ha^{-1} yr^{-1}	N Leached Nitrate kg ha^{-1} yr^{-1}	Reduction[c] Total %	Reduction[c] Nitrate %
Fallow	None	Loamy sand	CT[1]	112	127	36		
	Turnips				16	6	87	83
Fallow	None	Sand & clay	Sweden[2]	375	197	41		
	Rape				69	16	65	61
Lespedeza	None	Silt loam	KY[3]	224	65	18		
	Rye				17	5	74	72
Sudangrass	None	Loam	CA[4]	112	84	52		
	Mustard				17	10	80	81
	Purple vetch				75	36	11	31
	Sweet clover				83	43	1	17
Tobacco	None	Sandy loam	CT[5]	224	83	24		
	Oats				36	12	57	50
	Rye				28	9	66	62
	Timothy				57	16	31	33
Winter wheat	None	Silt	France[6]	200	110	54		
	Ryegrass				40	28	64	48

[a]Adapted from Meisinger et al., 1991.
[b]Reference for each study location is: 1, Volk and Bell, 1945; 2, Bertilsson, 1988; 3, Karraker et al., 1950; 4, Chapman et al., 1949; 5, Morgan et al., 1942; and 6, Martinez and Guirard, 1990.
[c]Percent reduction in N leached due to cover crops.

fectively as nonlegumes (Table 5). Cover crops can reduce nitrate leaching by 20 to 80% (Meisinger et al., 1991), with nonlegumes being about two to three times more effective than legumes for reducing leaching.

For studies summarized in Table 6, inclusion of cover crops reduced runoff 50% due to increased infiltration. It also reduced erosion 85%, total N loss 80%, and total P loss 71% due to vegetative protection of the surface soil. However, cover crops less effectively reduced nitrate and dissolved P losses (average 61 and 37%, respectively), mainly due to increased nitrate and dissolved P concentrations in runoff because the cover crops decreased runoff volume.

Research on Soil Management Techniques

Water Conservation

Past Needs

Although the need for water by plants has long been recognized, strong emphasis on storing (conserving) water in soil for later use by plants began mainly in the first half of this century. Research to conserve water and, therefore, to improve crop growth under dryland conditions was initiated at many locations after the major drought and associated dust storms that plagued the U.S. Great Plains states, Canadian prairie provinces, and surrounding regions during the 1930s. Early research showed that the same practices that provided protection against erosion by wind or water also were important for soil water conservation. Crop residue maintenance on the soil surface was found to be highly effective for these purposes. Consequently, soil and water conservation often have been investigated simultaneously. Through effective water conservation along with soil conservation, the vast USA Great Plains region now is an important agricultural region. Early water conservation efforts also helped improve crop production in more humid regions where short-term droughts can greatly reduce yields.

Current Status

Whereas much of the past research focused on practices that conserve water, current research often is focused on obtaining an understanding on the mechanisms through which the water conservation is achieved. Through such understanding, it should be possible to improve previously developed practices and to develop even more effective practices for conserving water.

Besides conserving water for crop production, it is highly important that the agricultural community act responsibly by using water available to it efficiently because the amount available often is limited and other users often compete for the same water. Therefore, besides research on water conserva-

Table 6. Effect of Cover Crops on Soil, N, and P Loss for Several Management Systems

			Fertilizer			Soil	N Loss		P Loss	
Crop	Cover	Loc.[a]	N kg ha^{-1} yr^{-1}	P kg ha^{-1} yr^{-1}	Runoff, mm	Loss kg ha^{-1} yr^{-1}	Nitrate kg ha^{-1} yr^{-1}	Total kg ha^{-1} yr^{-1}	Dissolved kg ha^{-1} yr^{-1}	Total kg ha^{-1} yr^{-1}
Corn	None	MD[1]	67	47	4	262	0.36	0.95	0.02	0.14
	Barley		67	47	1	33	0.05	0.12	0.01	0.01
Corn	None	GA[2]	—	20	159	3663	—	—	0.28	4.08
	Winter rye		—	50	97	938	—	—	0.30	1.39
Cotton	None	AL[3]	101	0	91	1067	1.40	3.13	0.31	0.44
	Winter wheat		101	0	35	260	0.56	0.88	0.16	0.20
Soybean	None	MO[4]	15	12	231	1439	3.36	—	0.46	—
	Common chickweed		15	12	133	233	0.77	—	0.17	—
	Canada bluegrass		15	12	142	93	0.88	—	0.43	—
	Downy brome		15	12	116	118	0.84	—	0.27	—
Wheat	None	NY[5]	241	64	173	—	1.14	—	0.32	—
	Ryegrass-alfalfa		241	64	74	—	0.93	—	0.17	—

[a]Reference for each study location is: 1, Angle et al., 1984; 2, Langdale et al., 1985; 3, Yoo et al., 1988; 4, Zhu et al., 1989; and 5, Klausner et al., 1974.

tion per se, the research often is aimed at improving water-use efficiency for crop production. Lysimeter facilities such as those at Coshocton, Ohio, and Bushland, Texas (Schneider et al., 1993), and waterflow models (Bristow et al., 1986; Hammel et al., 1981; Steiner et al., 1991) are being used increasingly to improve our understanding regarding the complex interrelationships among soil, plant, air, and climatic factors or conditions.

Future Challenges

Competition for available water resources undoubtedly will increase among the different sectors of society. Within agriculture, water conservation will remain important, and efficient water use will become increasingly important. Research to accomplish water conservation and efficient use of water will involve traditional field studies, models, lysimeters, and other equipment and facilities.

Besides conducting the research, results of the research must be conveyed to farmers through an effective technology transfer system. Even now, effective practices often are not used by farmers because of limited or ineffective technology transfer activities. While agencies other than research generally are responsible for technology transfer, a closer relationship between research and other groups may be required to achieve wider acceptance of effective water conservation and use practices.

Water Quality

Past Needs

Soil-management research for water quality has focused on developing practices that reduce N, P, and pesticide losses in ground- and surface waters, while maintaining optimum crop yield goals. This research was driven by the need to define the role of agriculture in nonpoint source pollution of water resources. In the 1970s, industrial, municipal, and urban sources of pollution were identified as major contributors to degrading water quality. Since passage of the Clean Water Act in 1972, much progress has been made in controlling pollution from point sources. As further control of pollution from remaining point sources becomes increasingly less cost-effective, and as significant water quality problems remain unresolved, more attention is being directed toward controlling pollution from agricultural nonpoint sources.

Current Status

Past research showed that general soil-management modifications over broad areas sometimes have not achieved expected water quality improvements. In fact, control measures are much more effective if concentrated on specific source areas rather than on entire watersheds (Heatwole et al., 1987; Prato and Wu, 1991). Therefore, current soil-management research directed at water quality is focusing on developing techniques that will allow us to identify areas that are major contributors of N or P to water resources, especially for watersheds with intensive confined animal operations, which often produce more N and P in manure than local crop requirements.

Future Challenges

With current technologies, chemical amendments are essential to maintain optimum crop yields that meet both local and world needs. However, if nutrient input rates are greater than crop removal rates, accumulations can occur in soil, which can increase the potential for chemical losses. Present water quality research often deals with one chemical. Future research should emphasize integrated programs involving C, N, and P. It should also consider holistic watershed management by balancing inputs and outputs on a large scale rather than at a field level. We must look beyond pure soil science and agronomic research and involve scientists from other disciplines. Most importantly, we must consider the economic impact of any changes in soil management on agricultural or rural communities.

Realistic water quality criteria. Water quality criteria for N and P have been established by the USEPA (1976). However, there is ongoing debate on whether to use maximum daily loadings or concentrations as the basis for management recommendations. Much of this debate centers on P due to a major difference between critical P values in water for eutrophication (0.01 to 0.02 mg L^{-1}) and in soil for crop growth (0.20 to 0.30 mg L^{-1}) as illustrated in Figure 4. This disparity emphasizes the sensitivity of many waters to inputs of P from agriculture.

Water quality criteria should not be used as the sole criteria to guide soil management where N and P losses are of concern. A more flexible approach considers relationships among N and P loadings, watershed characteristics, and use of the affected water. Such approach should encompass more than just N and P concentrations in runoff from impacted fields because unrealistic or unattainable criteria will not be adopted unless regulated. Phasing in of environmentally-sound management policies may receive wider acceptance and compliance by farmers without creating severe economic hardships within rural communities.

Economic and environmental sustainability. Sustainable soil management must involve agronomic, economic, and environmental compromises. For example, it may be necessary to apply P below the soil surface to reduce P in runoff and to periodically plow no-tillage soils to redistribute N and P throughout the plow layer. Both practices may indirectly reduce P loss by increasing crop uptake of P and yield, which affords more vegetative protection

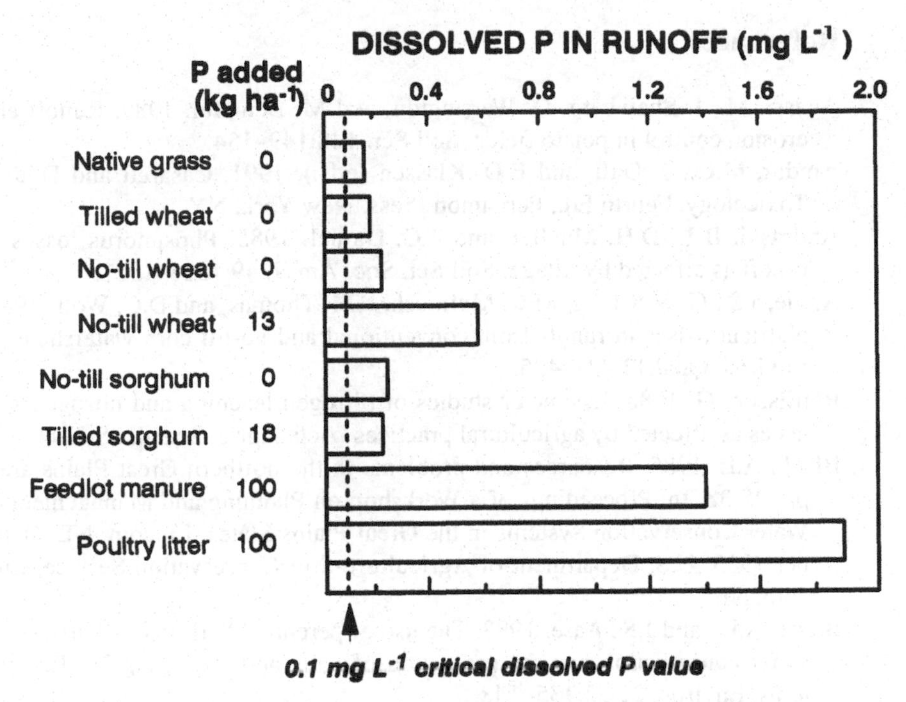

Figure 4. The mean flow-weighted concentration of dissolved P in runoff during one year as a function of watershed and manure management in the Southern Plains relative to critical values associated with accelerated eutrophication (adapted from Jones et al., 1995; Heathman et al., 1995).

of surface soil from erosion. However, conflicts may exist among best management practices (BMPs), residue management guidelines, and recommended subsurface applications. Thus, BMPs should be flexible enough to enable modified residue and nutrient management plans to be compatible.

Technology transfer and education programs. Although we have reduced water pollution due to point sources of N and P, less has been done regarding pollution due to nonpoint agricultural sources. To achieve this, we must identify critical sources in a watershed to target cost-effective remedial strategies. Perhaps most crucial to any water quality improvement strategy is efficient transfer of technology to the farmer. This will involve education programs to overcome the perception by water users that it is often cheaper to treat the symptoms of degradation than to control the nonpoint sources. Unfortunately, benefits of such programs often are not immediately visible to a concerned public. Future research and education programs should emphasize the long-term economic and environmental benefits of these measures.

References

Agassi, M., I. Shainberg, D. Warrington, and M. Ben-Hur. 1989. Runoff and erosion control in potato fields. Soil Sci. 148:149–154.

Amdur, M.O., J. Dull, and E.D. Klassen (eds.). 1991. Casarett and Doull's Toxicology. Fourth Ed., Pergamon Press, New York, NY.

Andraski, B.J., D.H. Mueller, and T.C. Daniel. 1985. Phosphorus losses in runoff as affected by tillage. Soil Sci. Soc. Am. J. 49:1523–1527.

Angle, J.S., G. McClung, M.C. McIntosh, P.M. Thomas, and D.C. Wolf. 1984. Nutrient losses in runoff from conventional and no-till corn watersheds. J. Environ. Qual. 13:431–435.

Bertilsson, G. 1988. Lysimeter studies of nitrogen leaching and nitrogen balances as affected by agricultural practices. Acta Agric. Scand. 38:3–11.

Black, A.L. 1985. Resources and problems in the northern Great Plains area. pp. 25-38. In: Proceedings of a Workshop on Planning and Management of Water Conservation Systems in the Great Plains States, Lincoln, NE, October 1985. U.S. Department of Agriculture-Soil Conservation Service, Lincoln, NE.

Black, A.L., and J.K. Aase. 1988. The use of perennial herbaceous barriers for water conservation and the protection of soils and crops. Agric., Ecosystems, Environ. 22/23:135–148.

Black, A.L., and F.H. Siddoway. 1977. Winter wheat recropping on dryland as affected by stubble height and nitrogen fertilization. Soil Sci. Soc. Am. J. 41:1186–1190.

Bristow, K.L., G.S. Campbell, R.I. Papendick, and L.F. Elliott. 1986. Simulation of heat and moisture transfer through a surface residue-soil system. Agric. Forest Meteorol. 36:193–214.

Bruce, R.R., G.W. Langdale, L.T. West., and W.P. Miller. 1995. Surface soil degradation and soil productivity restoration and maintenance. Soil Sci. Soc. Am. J. 59:654–660.

Burnett, E., and V.L. Hauser. 1967. Deep tillage and soil-plant-water relationships. pp. 47–52. In: Proceedings of Tillage for Greater Crop Production Conference, December 1967. American Society of Agricultural Engineers, St. Joseph, MI.

Burwell, R.E., R.R. Allmaras, and L.L. Sloneker. 1966. Structural alteration of soil surfaces by tillage and rainfall. J. Soil Water Conserv. 21:61–63.

Burwell, R.E., D.R. Timmons, and R.F. Holt. 1975. Nutrient transport in surface runoff as influenced by soil cover and seasonal periods. Soil Sci. Soc. Am. Proc. 39:523–528.

Carter, D.L., R.D. Berg, and B.J. Sanders. 1991. Producing no-till cereal or corn following alfalfa on furrow irrigated land. J. Prod. Agric. 4:174–179.

Chapman, H.D., G.F. Liebig, and D.S. Rayner. 1949. A lysimeter investigation of nitrogen gains and losses under various systems of covercropping and fertilization and a discussion of error sources. Hilgardia 19:57–95.

Charreau, C. 1977. Some controversial technical aspects of farming systems in semi-arid west Africa. pp. 313-360. In: G.H. Cannell (ed.) Proceedings of International Symposium on Rainfed Agriculture in Semi-Arid Regions, Riverside, CA, April 1977. University of California, Riverside.

CTIC (Conservation Technology Information Center). 1990. Tillage definitions. Conserv. Impact 8(10):7.

Eck, H.V., and H.M. Taylor. 1969. Profile modification of a slowly permeable soil. Soil Sci. Soc. Am. Proc. 33:779–783.

Edwards, D.R., T.C. Daniels, and O. Marbun. 1992. Determination of best timing for poultry waste disposal: A modeling approach. Water Resour. Bull. 28:487–494.

Edwards, W.M., L.D. Norton, and C.E. Redmond. 1988a. Characterizing macropores that affect infiltration into nontilled soil. Soil Sci. Soc. Am. J. 52:483–487.

Edwards, W.M., M.J. Shipitalo, and L.D. Norton. 1988b. Contribution of macroporosity to infiltration into a continuous corn no-tilled watershed: Implications for contaminant movement. J. Contam. Hydrol. 3:193–205.

Follett, R.F., S.C. Gupta, and P.G. Hunt. 1987. Conservation practices: Relation to the management of plant nutrients for crop production. pp. 19-51. In: R.F. Follett et al. (ed.), Soil fertility and organic matter as critical components of production systems. SSSA Special Publication Number 19, Soil Science Society of America, Madison, WI.

Griffith, D.R., J.V. Mannering, and W.C. Moldenhauer. 1977. Conservation tillage in the eastern Corn Belt. J. Soil Water Conserv. 32:20–28.

Guertal, E.A., D.J. Eckert, S.J. Traina, and T.J. Logan. 1991. Differential phosphorus retention in soil profiles under no-till crop production. Soil Sci. Soc. Am. J. 55:410–413.

Hammel, J.E., R.I. Papendick, and G.S. Campbell. 1981. Fallow tillage effects on evaporation and seedzone water content in a dry summer climate. Soil Sci. Soc. Am. J. 45:1016–1022.

Harper, H.J. 1941. The effect of terrace ridges on the production of winter wheat. Soil Sci. Soc. Am. Proc. 6:474–479.

Heathman, G.C., A.N. Sharpley, S.J. Smith, and J.S. Robinson. 1995. Poultry litter application and water quality in Oklahoma. Fert. Res. 40:165–173.

Heatwole, C.D., A.B. Bottcher, and L.B. Baldwin. 1987. Modeling cost-effectiveness of agricultural nonpoint pollution abatement programs on two Florida basins. Water Resour. Bull. 23:127–131.

Jacks, G.V., W.D. Brind, and R. Smith. 1955. Mulching. Technical Communication Number 49, Commonwealth Bureau of Soil Science (England).

Jones, O.R. 1981. Land forming effects on dryland sorghum production in the Southern Great Plains. Soil Sci. Soc. Am. J. 45:606–611.

Jones, O.R., and R.N. Clark. 1987. Effects of furrow dikes on water conservation and dryland crop yields. Soil Sci. Soc. Am. J. 51:1307–1314.

Jones, O.R., and V.L. Hauser. 1975. Runoff utilization for grain production. pp. 277-283. In: Proceedings of a Water Harvesting Symposium, Phoenix,

AZ, March 1974. USDA-ARS-W-22. U.S. Government Printing Office, Washington, D.C.

Jones, O.R., V.L. Hauser, and T.W. Popham. 1994. No-tillage effects on infiltration, runoff, and water conservation on dryland. Trans. Am. Soc. Agric. Eng. 37:473–479.

Jones, O.R., W.M. Willis, S.J. Smith, and B.A. Stewart. 1995. Nutrient cycling of cattle feedlot manure and composted manure applied to Southern High Plains drylands. pp. 265-272. In: K. Steele (ed.) Impact of Animal Manure and the Land-Water Interface. Lewis Publishers, Boca Raton, FL.

Kanwar, R.S., J.L. Baker, and J.M. Laflen. 1985. Nitrate movement through the soil profile in relation to tillage systems and fertilizer application methods. Trans. Am. Soc. Agric. Eng. 28:1802–1807.

Karraker, P.E., C.E. Bortner, and N.E. Fergus. 1950. Nitrogen balance in lysimeters as affected by growing Kentucky bluegrass and certain legumes separately and together. Bulletin Number 557, Kentucky Agricultural Experiment Station, Lexington.

Klausner, S.D., P.J. Zwerman, and D.F. Ellis. 1974. Surface runoff losses of soluble nitrogen and phosphorus under two systems of soil management. J. Environ. Qual. 3:42–46.

Langdale, G.W., R.A. Leonard, and A.W. Thomas. 1985. Conservation practice effects on phosphorus losses from Southern Piedmont watersheds. J.Soil Water Conserv. 40:157–160.

Lentz, R.D., I. Shainberg, R.E. Sojka, and D.L. Carter. 1992. Preventing irrigation furrow erosion with small applications of polymers. Soil Sci. Soc. Am. J. 56:1926–1932.

Lentz, R.D., and R.E. Sojka. 1994. Field results using polyacrylamide to manage furrow erosion and infiltration. Soil Sci. 158:274–282.

Lyle, W.M., and J.P. Bordovsky. 1981. Low energy precision application (LEPA) irrigation system. Trans. Am. Soc. Agric. Eng. 24:1241–1245.

Martinez, J., and G. Guirard. 1990. A lysimeter study of the effects of a ryegrass catch crop, during a winter wheat/maize rotation, on nitrate leaching and on the following crop. J. Soil Sci. 41:5–16.

Meisinger, J.J., W.L. Hargrove, R.L. Mikkelsen, J.R. Williams, and V.W. Benson. 1991. Effects of cover crops on groundwater quality. pp. 57–68. In: W.L. Hargrove (ed.) Cover crops for clean water. Soil and Water Conservation Society, Ankeny, IA.

Morgan, M.F., H.G.M. Jacobson, and S.B. LeCompte, Jr. 1942. Drainage water losses from a sandy soil as affected by cropping and cover crops. Bulletin 466, Connecticut Agricultural Experiment Station, New Haven.

Mostaghimi, S., T.A. Dillaha, and V.O. Shanholtz. 1988. Influence of tillage systems and residue level on runoff, sediment, and phosphorus losses. Trans. Am. Soc. Agric. Eng. 31:128–132.

Mostaghimi, S., T.M. Younos, and U.S. Tim. 1992. Crop residue effects on nitrogen yield in water and sediment runoff from two tillage systems. Agric., Ecosystems, Environ. 39:187–196.

Musick, J.T. 1970. Effect of antecedent soil water on preseason rainfall storage in a slowly permeable irrigated soil. J. Soil Water Conserv. 25:99–101.

OECD (Organization for Economic Cooperation and Development). 1982. Eutrophication of Waters: Monitoring, Assessment, and Control. OECD, Paris, France.

Olson, R.J., R.F. Hensler, O.J. Attoe, S.A. Witzel, and L.A. Peterson. 1970. Fertilizer nitrogen and crop rotation in relation to movement of nitrate nitrogen through soil profiles. Soil Sci. Soc. Am. Proc. 34:448–452.

Papendick, R.I., M.J. Lindstrom, and V.L. Cochran. 1973. Soil mulch effects on seedbed temperature and water during fallow in Eastern Washington. Soil Sci. Soc. Am. Proc. 37:307–313.

Papendick, R.I., and D.E. Miller. 1977. Conservation tillage in the Pacific Northwest. J. Soil Water Conserv. 32:49–56.

Power, J.F., W.W. Wilhelm, and J.W. Doran. 1986. Crop residue effects on soil environment and dryland maize and soybean production. Soil Tillage Res. 8:101–111.

Prato, T., and S. Wu. 1991. Erosion, sediment and economic effects of conservation compliance in an agricultural watershed. J. Soil Water Conserv. 46:211–214.

Richardson, C.W. 1973. Runoff, erosion, and tillage efficiency on graded-furrow and terraced watersheds. J. Soil Water Conserv. 28:162–164.

Rosenberg, N.J. 1966. Microclimate, air mixing and physiological regulation of transpiration as influenced by wind shelter in an irrigated bean field. Agric. Meteorol. 3:197–224.

Schneider, A.D., T.A. Howell, and J.L. Steiner. 1993. An evapotranspiration research facility using monolithic lysimeters from three soils. Appl. Eng. Agric. 9:227–235.

SCSA [Soil Conservation Society of America (now SWCS, Soil and Water Conservation Society)]. 1976. Resource conservation glossary. Soil and Water Conservation Society, Ankeny, IA.

Sharpley, A.N. 1981. The contribution of phosphorus leached from crop canopy to losses in surface runoff. J. Environ. Qual. 10:160–165.

Sharpley, A.N., S.C. Chapra, R. Wedepohl, J.T. Sims, T.C. Daniel, and K.R. Reddy. 1994. Managing agricultural phosphorus for protection of surface waters: Issues and options. J. Environ. Qual. 23:437–451.

Sharpley, A.N., J.J. Meisinger, A. Breeuwsma, J.T. Sims, T.C. Daniel, and J.S. Schepers. 1995. Impacts of animal manure management on ground and surface water quality. In: J. Hatfield (ed.) Effective management of animal waste as a soil resource. John Wiley, New York (in press).

Sharpley, A.N., J.J. Meisinger, J. Power, and D.L. Suarez. 1992a. Root extraction of nutrients associated with long-term soil management. Adv. Soil Sci. 19:151–217.

Sharpley, A.N., and S.J. Smith. 1989. Mineralization and leaching of phosphorus from soil incubated with surface-applied and incorporated crop residues. J. Environ.Qual. 18:101–105.

Sharpley, A.N., and S.J. Smith. 1991. Effects of cover crops on surface water quality. pp. 41–49. In: W.L. Hargrove (ed.) Cover crops for clean water. Soil and Water Conservation Society, Ankeny, IA.

Sharpley, A.N., and S.J. Smith. 1993. Wheat tillage and water quality in the Southern Plains. Soil Tillage Res. 30:33–48.

Sharpley, A.N., S.J. Smith, O.R. Jones, W.A. Berg, and G.A. Coleman. 1992b. The transport of bioavailable phosphorus in agricultural runoff. J. Environ. Qual. 21:30–35.

Sharpley, A.N., S.J. Smith, J.R. Williams, O.R. Jones, and G.A. Coleman. 1991. Water quality impacts associated with sorghum culture in the Southern Plains. J. Environ. Qual. 20:239–244.

Shipitalo, M.J., W.M. Edwards, W.A. Dick, and L.B. Owens. 1990. Initial storm effects on macropore transport of surface-applied chemicals in no-till soil. Soil Sci. Soc. Am. J. 54:1530–1536.

Smika, D.E., and G.A. Wicks. 1968. Soil water storage during fallow in the Central Great Plains as influenced by tillage and herbicide treatments. Soil Sci. Soc. Am. Proc. 32:591–595.

Smith, S.J., J.W. Naney, and W.A. Berg. 1987. Nitrogen and ground water protection. pp. 367-374. In: D.M. Fairchild (ed.) Ground Water Quality and Agricultural Practices. Lewis Publishers, Chelsea, MI.

Smith, S.J., A.N. Sharpley, J.W. Naney, W.A. Berg, and O.R. Jones. 1991. Water quality impacts associated with wheat culture in the Southern Plains. J. Environ. Qual. 20:244–249.

Sprague, M.A. 1979. Agronomy. pp. 27-29. In: R.W. Fairbridge and C.W. Finkl, Jr. (eds.) The Encyclopedia of Soil Science, Part I, Physics, Chemistry, Biology, Fertility and Technology. Dowden, Hutchinson & Ross, Stroudsburg, PA.

SSSA (Soil Science Society of America). 1987. Glossary of Soil Science Terms. Soil Science Society of America, Madison, WI.

Staple, W.J., J.J. Lehane, and A. Wenhardt. 1960. Conservation of soil moisture from fall and winter precipitation. Can. J. Soil Sci. 40:80–88.

Staricka, J.A., R.R. Allmaras, and W.W. Nelson. 1991. Spatial variation of crop residue incorporated by tillage. Soil Sci. Soc. Am. J. 55:1668–1674.

Steiner, J.L. 1989. Tillage and surface residue effects on evaporation from soils. Soil Sci. Soc. Am. J. 53:911–916.

Steiner, J.L., T.A. Howell, J.A. Tolk, and A.D. Schneider. 1991. Evapotranspiration and growth predictions of CERES-maize, -sorghum, and -wheat for water management in the southern High Plains. pp. 297–303. In: Proceedings American Society of Civil Engineers National Conference on Irrigation and Drainage Research, Honolulu, HI, July 1991.

Stewart, B.A., and W.C. Moldenhauer (eds.). 1994. Crop residue management to reduce erosion and improve soil quality: Southern Great Plains. Conservation Research Report Number 37, U.S. Department of Agriculture-Agricultural Research Service.

Stewart, B.A., and J.L. Steiner. 1990. Water-use efficiency. Adv. Soil Sci. 13:151–173.

Stewart, B.A., D.A. Woolhiser, W.H. Wischmeier, J.H. Caro, and M.H. Frere. 1975. Control of water pollution from cropland. Volume 1. A manual for guideline development. U.S. Department of Agriculture Report Number ARS-H-5-1. U.S. Government Printing Office, Washington, D.C.

Triplett, Jr., G.B., and D.M. Van Doren, Jr. 1977. Agriculture without tillage. Sci. Am. 236(1):28–33.

Tyler, D.D., and G.W. Thomas. 1977. Lysimeter measurements of nitrate and chloride losses from conventional and no-tillage corn. J. Environ.Qual. 6:63–66.

Unger, P.W. 1984. Tillage and residue effects on wheat, sorghum, and sunflower grown in rotation. Soil Sci. Soc. Am. J. 48:885–891.

Unger, P.W. 1992. Infiltration of simulated rainfall: Tillage system and crop residue effects. Soil Sci. Soc. Am. J. 56:283–289.

USEPA (U.S. Environmental Protection Agency). 1976. Quality criteria for water. U.S. Government Printing Office, Washington, D.C.

Van Doren, Jr., D.M., and R.R. Allmaras. 1978. Effect of residue management practices on the soil physical environment, microclimate, and plant growth. pp. 49-83. In: W.R. Oschwald (ed.) Crop residue management systems. ASA Special Publication Number 31, American Society of Agronomy, Madison, WI.

Volk, G.M., and C.E. Bell. 1945. Some major factors in the leaching of calcium, potassium, sulfur and nitrogen from sandy soils: A lysimeter study. Bulletin Number 416, University of Florida, Gainesville.

Yoo, K.H., J.T. Touchton, and R.H. Walker. 1988. Runoff, sediment and nutrient losses from various tillage systems of cotton. Soil Tillage Res. 12:13–24.

Zhu, J.C., C.J. Gantzer, S.H. Anderson, E.E. Alberts, and P.R. Beuselinck. 1989. Runoff, soil, and dissolved nutrient losses from no-till soybean with winter cover crops. Soil Sci. Soc. Am. J. 53:1210–1214.

Zingg, A.W., and V.L. Hauser. 1959. Terrace benching to save potential runoff for semiarid land. Agron. J. 51:289–292.

Zingg, A.W., and C.J. Whitfield. 1957. A summary of research experience with stubble-mulch farming in the Western States. U.S. Department of Agriculture Technical Bulletin Number 1166. U.S. Government Printing Office, Washington, D.C.

Agricultural Technology and Adoption of Conservation Practices

R.R. Allmaras, D.E. Wilkins, O.C. Burnside, and D.J. Mulla

INTRODUCTION

Technology inputs from research have transformed American agriculture in the last 60 years from a resource-based industry to a science-oriented industry, or from a traditional to a high technology sector (Ruttan, 1982). Three inputs of equal importance were responsible for technology adoption and the transformation. These are: (a) the capacity of public- and private-sector research to produce the new technical knowledge; (b) the capacity of the industrial sector to develop, produce, and market new technology; (c) and the capacity of farmers to acquire new knowledge and use inputs effectively. In his historical account of agriculture development in the United States, Cochrane (1993) notes that economic survival through reduced input costs and more production per unit cost has always been and still is a concern of most farmers. Factors external to the farm have changed somewhat over time but demand for agricultural produce has generally lagged behind production. In these accounts of agricultural development in the United States, how is it that soil and water conservation might receive attention? Where and when does scientific development of conservation practices fit into this transformation of American agriculture through application of technology?

Soil and water conservation controls environment both within the field and off-site; control may be achieved through either management or structures (support practices). Structures inputs are terraces, grassed waterways, terrace outlets, tiles or surface waterways for drainage, divided slopes, contour strips, grass barriers for snow catch, shelterbelts, and riparian strips. Management inputs are tillage systems, crop residue management, crop rotations, methods of pest control, traffic control, and methods and systems to enhance the production of crop residues. These inputs can be dramatic or subtle for control of water infiltration, soil erosion, internal drainage, soil/water/air quality, and soil productivity.

The last 60-year period provides an excellent evaluation for the progress of soil and water conservation for crop production in the United States. Before 1940, agricultural production was dependent upon increasing the land under cultivation, but since 1940 agricultural production has expanded by technology applications to a shrinking land base. Scientific activity needed for technology innovations in agriculture was under development somewhat intensively since about 1890. A recognition of these major shifts in American agriculture is needed to understand the linkages of science, technology, and conservation of soil and water. Although there may be obvious trade-offs between crop production and conservation of soil and water resources, as discussed by Heady (1984), there have been incidental benefits and damages as well, to soil and water conservation from the adoption of technology for crop production. Crop residue management is an incidental benefit to soil quality that may have prevented a continued environment of soil and water degradation such as emerged in the 1930 era of the Dust Bowl. Perhaps the most pressing problem associated with modern agricultural technology is nonpoint source pollution.

Our discussion briefly addresses the following objectives:

1. Characterize soil and water conservation practices and changes during the 60-year period—benefits and problems associated with the structures and management types of input delineated above.
2. Characterize technology inputs and changes on the land over the past 60 years and the impact of scientific research and development on crop production inputs and changes, as well as on soil and water conservation.
3. Discuss linkages between technology and soil/water conservation to evaluate progress and to project emerging issues in the production of food and fiber.

The Merging of Science, Conservation, and Technology (pre-1940)

Historical developments late in the nineteenth and the first 40 years of the twentieth century were a necessary precursor to the technology revolution after 1940. Events leading up to the 1940 landmark passed slowly and were often difficult for agriculture. Cochrane (1993) states emphatically that agricultural development in the 1940 to 1990 era cannot be understood and appreciated without a historical perspective over the four centuries of agricultural activity in the United States. Nearly all lands except for irrigated tracts had been tilled for crop production by 1900, hence there was little expansion into new lands after 1900. The regional specialization of crop systems and commodity production depicted by the Major Land Resource Regions (Austin, 1981) was already developed, but regional specialization since 1940 has intensified somewhat because of the general specialization to one or two crops instead of the original emphasis on crop rotation (Heady, 1984). Notable changes since 1900 were the shift in cotton production from southeastern United States to the Southern Plains and the irrigated West. Soybean production in the Corn Belt and Southeast has also been developed since 1920.

Late in the nineteenth century crop production increased markedly while yields changed little (Table 1, see crops and botanical names), because new

Table 1. Yields of Ten Crops for Selected Years During the Period from 1870 to 1990[a]

Year	Barley kg/ha	Oat kg/ha	Corn kg/ha	Sorghum kg/ha	Soybean kg/ha	Sunflower kg/ha	Wheat kg/ha	Cotton kg/ha	Potato t/ha	Hay t/ha
1870	1200	1010	—[b]	—	—	—	870	200	6.0	2.6
1900	1310	1080	1590	—	—	—	900	210	5.7	2.8
1930	1140	1090	1500	880	940	—	990	200	7.5	2.6
1935	1070	880	1190	670	1050	—	840	210	7.6	2.5
1940	1280	1150	1890	930	1260	—	1050	280	8.8	2.9
1945	1340	1220	2180	1080	1310	—	1180	300	11.1	3.1
1950	1430	1260	2380	1370	1480	—	1070	310	16.5	3.2
1955	1560	1310	2650	1960	1410	—	1330	440	19.9	3.4
1960	1640	1500	3540	2540	1650	—	1640	510	21.4	3.9
1965	2170	1680	4470	3170	1660	—	1790	580	23.2	4.2
1970	2430	1930	5240	3370	1860	1099	2180	500	25.8	4.8
1975	2380	1730	5240	3050	1790	1190	2010	520	29.0	4.8
1980	2790	1960	6580	3680	2020	1350	2330	570	30.9	5.3
1985	2830	2160	7320	4060	2180	1330	2530	680	33.2	5.6
1990	2920	2010	7310	3790	2290	1360	2430	720	33.7	5.4

[a]From Cochrane, 1993; Century of Agriculture in Charts and Tables, Agric. Hdbk. 318, U.S. Dept. Agric. Statistical Reporting Service, 1966; and Agricultural Statistics, USDA, 1956, 1972, 1973, 1978, 1983, 1988, and 1993. Three-year average centered on the year indicated. Crops with botanical names are: barley (*Hordeum vulgare* L.), oat (*Avena sativa* L.), corn (*Zea mays* L.), sorghum (*Sorghum bicolor* L.), soybean (*Glycine max* L. Merrill), sunflower (*Helianthus annus* L), wheat (*Triticum spp.*), cotton (*Gossypium hirsutem* L.), potato (*Solanum tuberosum* L.).
[b]No data reported.

lands were being put under cultivation. Some of these lands in the Great Plains were marginal because of drought when annual precipitation was less than 45 cm. Farmers in the Great Plains had not yet learned to cope with moisture deficit. Technology adoption during this period was mainly mechanical—row crop cultivators, forage mowers, reapers, sulky plows, spring tooth harrows, disk harrows, seed drills, and twine binders all powered by horses and mules. Stationary steam engines and threshing machines were beginning to appear. During the first 30 years of the twentieth century, output per unit input and crop yields showed no significant changes (Table 1), but total production increased at a steady rate of about 1% per year. Land area under cultivation in 1930 was about 200 million ha; now it is less than 160 million ha. Cochrane (1993) attributes these changes to mechanization including the gasoline tractor and larger or more sturdy versions of the machines under development since about 1860. The threshing machine became larger and thus crop residues were removed from the field and placed into straw piles that were often burned. The combine was also under development but was not used widely yet except for winter wheat harvest in the Pacific Northwest and California. There was little use yet of nitrogen fertilizers and lime, but manures were used extensively. Most of the crop improvement was from introduced cultivars. Another factor during the period from 1900 to 1933 was a chronic depression—farmers generally were not able to realize profits and often could not retire outstanding loans on land purchase and improvements. Chattel loans for seed and fuel were common.

Beginning in 1940 there were major improvements in crop yield (Table 1) which grew throughout the 1940 to 1990 era. Not only was mechanization improved, but biological (improved cultivars) and chemical inputs (fertilizers, lime, and pesticides) became available. These biological and chemical inputs required machinery and mechanization improvements, which could not have been accomplished without the mechanical technology that was already developed before 1940. It was during the late nineteenth and early twentieth centuries when science was linked to agricultural technology. This linkage of science and technology was initiated when the United States Department of Agriculture (USDA) was established in 1862 and was to become a science producing agency of the government; it was only after 1889 that the USDA also became an action agency to deal with economic and market problems of agriculture. Under the leadership of James Wilson (1897-1913) the USDA was vigorously transformed into a science institution. Details of such governmental and political activity are summarized by Larson et al. (1998). During almost the same period, agricultural colleges transformed agricultural technology through advances in plant breeding, plant disease, soil chemistry, and animal diseases. While both the USDA research bureaus and the agricultural colleges accelerated the development of agricultural technology in the first two decades of the twentieth century, the delivery system of the agricultural colleges (Extension Service) was readied for the era of scientific agriculture that began in 1940.

Prompted by destruction of forests late in the nineteenth century, the first formal linkage of science and conservation was made when Theodore Roosevelt decided in 1901 that conservation should not be politically managed, but rather it was to be a scientific endeavor. The United States Forest Service under Gifford Pinchot began scientific forest management. During the presidency of Franklin Roosevelt, the Soil Conservation Service (soil and water conservation research to control erosion) was formed and headed by Hugh Hammond Bennett in 1933 after both he and the president had witnessed the ravishes of water erosion on cropland early in the twentieth century. However, wind erosion and the dust storms of the early 1930 era literally frightened Americans into action to protect the environment.

Some typical crop rotations along with the benefits late in the pre-1940 era were discussed by Leighty (1938). After an acknowledgment that no one rotation can be typical and crop rotations in a farming region were built around a principal crop, Leighty (1938) discussed some crop rotation surveys. The principal crop of Leighty (1938) is similar to the hub crop (Pearson, 1967), which is the crop that offers the greatest comparative advantage. In the eastern Corn Belt the predominant 3-year rotation was corn, small grain (wheat or oat), and clover. Corn after corn without wheat was used to extend the rotations. In these rotations typical of Indiana, the residues were harvested for animal production, the primary tillage implement was a moldboard plow, and corn ears were picked and shelled off-site. In the subhumid to semiarid regions west of the Corn Belt, wheat (spring or winter types) was the hub crop (30 to 75% of the rotation) with about 25% fallow, sweet clover (*Melilotus spp.*) green manure, some corn, and up to 25% as feed grains (oat, barley) and flax (*Linum usitatissimum* L.). Sorghum replaced corn in the southern farming regions and alfalfa (*Medicago sativa* L.) was frequently grown for hay in the Corn Belt and eastern reaches of wheat country. Livestock production and the need for feeding draft animals provided diversification and the opportunity to use crop rotations for weed and disease control, plus control of erosion, because forage production was required to feed the draft animals. In these more humid wheat-producing areas, wheat residues were removed with binders and threshing machines, and much corn was harvested for silage. Primary tillage was predominantly moldboard plowing.

Crop rotation, tillage, and harvest methods were markedly different in the semiarid, dryland farming; wheat was generally alternated with summer fallow. Depending on soil-water-storage capability during the cold season, a green manure (a seeded cover crop or volunteer and weeds) preceded summer fallow, and wheat followed wheat, sometimes as a combination of winter and spring types. Late in the pre-1940 era the sweep (stubble mulch system) was being adopted, and headers were used instead of binders—combines instead of threshing machines. These practices were the beginning of more crop residue retention on the micro-site not only in the semiarid dryland but also in wheat and corn production of other regions.

Leighty (1938) provided an interesting discussion about the advantages of crop rotation. It used diversification (a) to protect against total failure, (b) to provide year-round distribution of income and work for men/machinery/horses, and (c) to facilitate a livestock enterprise at the farm level. Corn or wheat as a hub crop was rotated with legumes or other nonlegumes as a source of soil fertility. Leighty (1938) listed 24 plant diseases that were being controlled entirely or in part by crop rotation; however, numerous added recommendations were made to *plow under* or burn the diseased crop residue. Crop rotation or crop diversity (of growth pattern) was also discussed in detail for pest control. Again the caveat was the marked difference among farm operators in adjacent fields. The discussion also implied growing weed pressure due to careless seed contamination and lack of sanitation in the roadways and other infrastructures adjacent to farm fields. Leighty (1938) discussed the value of crop rotation for soil erosion control while using 14 years of runoff/erosion measurements on the Shelby loam in Missouri (a corn-wheat-clover rotation was compared with a cultivated bare soil, continuous wheat, and continuous corn).

While Leighty (1938) focused primarily on central and northern United States and did not discuss cover crops, Pieters and McKee (1938) discussed cover and green manure crops. Their focus on cover crops in southeastern United States included the control of soil erosion, which had already been measured from 1931 to 1935 in Statesville, North Carolina.

During the 1931-1942 period, there was much activity concerned with soil erosion in the semiarid wheat and pea (*Pisum sativum* L.) lands of the Pacific Northwest (Horner et al., 1944). Stubble mulch was not successful, green manure crops including sweet clover were hard to manage because of undesired seed dispersal and increased production costs, and soil erosion was estimated to average 20 tons/ha per year.

1940 to 1995: An Era of Intense Technology Adoption in Agriculture

This period contrasts sharply with the pre-1940 era, when American farmers were not adopting technology as rapidly because of poor economic conditions and a meager supply or development of technology. Most technology adoption before 1940 was of the mechanical or machinery type, whereas mechanical, biological, and chemical technologies became available beginning about 1940 to 1950. It was an era also when there were major improvements in soil and water conservation.

Abrupt changes in unit area production (Table 1) indicate big changes in technology adoption within the 1940 to 1995 era. Unit area production for nearly all crops (Table 1) was stagnated until 1940, when a large change occurred within a five-year period and growth until 1960 was modest. In 1960 there was another sharp increase over a five-year period followed by a larger increase rate than in the earlier 1940 to 1960 period. A third sharp increase can be seen in the 1975 to 1980 period, and crop yields have increased moderately since 1980.

The total output from American agriculture including animal enterprises over the 1948 to 1991 period increased 1.9%/yr while the total input decreased 0.05 %/yr; the multifactor productivity was 1.95%/yr. Total input was nearly constant (with a 1948 equivalent of 100) from 1948 until 1970 when it increased to an index of 110 in 1978, but since 1978 the index has dropped to 90 (ERS, 1994). This drop indicates a more efficient integration of mechanical, biological, and chemical technology as discussed by Cochrane (1993). Since 1980 the direct use of energy in agriculture (i.e., diesel fuel and gasoline) has declined, the use of fertilizers is at a plateau, and herbicide use has dropped (ERS, 1994). Purchased inputs of agricultural machinery have also declined nearly 30% since 1979, but have increased somewhat since 1990.

From 1940 to 1980 the biological technology factor of genetic improvement (Fehr, 1984) contributed about 40, 70, 80, 30, and 55% of the respective unit area productions of sorghum, corn, soybean, cotton, and wheat shown in Table 1. However, there were no obvious changes in these average genetic improvements to correspond to the large change in unit area yield increase noted in about 1960. Introduction and use of the chemical type technology (ERS, 1994) explains some of the unit area production and may explain part of the different increase rates in 1940 to 1960 compared with 1960 to 1980. The phosphate and potash fertilizers were introduced about 1940, their use increased at a constant rate until 1980, and then remained steady after 1980. The use of nitrogen (N) fertilizers began a sharp increase in about 1960 and remains constant since 1981 (ERS, 1994). Herbicide and insecticide use sharply increased beginning about 1960, but the use has declined somewhat after maximum use in about 1984 (ERS, 1994). Consequently, these significant increases in the use of N fertilizer and herbicide in 1960 may explain the curious break in unit area production between the 1940 to 1960 era and 1960 to 1980 era. Accountability of these factors in unit area production can only be approximate because of institutional changes such as a dramatic increase in farm size beginning in 1950 (Cochrane, 1993).

Not only was there a major change in adopted technology in the 1940 to 1990 era but also there were many changes/developments in soil and water conservation. Soil and water conservation research was increased, conservation planning was initiated, soil and water resource assessment became a national initiative, and new problems of environmental protection emerged (Larson et al., 1998). As discussed in an earlier section, Gifford Pinchot and John Muir had spearheaded public and governmental action to protect forestland, rangeland, and wildlife habitat before 1940. Recognition of soil erosion in arable croplands in the humid east, as well as in the semiarid and arid Great Plains prompted public and governmental action in 1935 with the appointment of Hugh Hammond Bennett as head of the Soil Conservation Service to initiate research and assessment of soil erosion nationwide (Miller et al., 1985). The first national assessment of soil erosion damage was completed in 1935, and later assessments were legislated in 1977 to appraise the soil and water resources of the country (Larson et al., 1998); other resource and soil erosion related activities are conservation planning and conservation compliance at the farm level.

A critically necessary conservation activity of this period is the research for soil erosion prediction technology. These research activities for understanding and controlling water and wind erosion, initiated by the Soil Conservation Service in about 1940, were assumed by the Agricultural Research Service and Experiment Stations in about 1950; these erosion prediction technologies are now used in all farm plans for conservation compliance, but are still being researched as farmers are adopting new technologies for land management and crop production.

Already in the 1940 to 1950 era, field research was in progress to understand specific needs for tillage aimed at reduced tillage, and ultimately no-till or direct drilling methods (Blevins et al., 1998). This tillage research demonstrated the benefits to soil and water conservation, and pioneered the three technologies (mechanical, biological, and chemical) needed to develop these new systems, yet it was already 1970 when conservation tillage was researched specifically to manage crop residues for soil and water conservation (Larson et al., 1998).

Surface water quality and other off-site damages from runoff and soil erosion, as well as leaching to groundwater, are now a major soil and water concern; environmental and ecological concerns for agriculture have also emerged. This environmental concern, raised when Rachel Carson published *Silent Spring* in 1962, has prompted many public and governmental reactions (Larson et al., 1998). Within arable agriculture, a whole new range of technologies were and are still being developed to understand the mechanisms of solute movement, reduce the amounts of agricultural chemicals used, control placement and timing of agrichemical application, and reduce surface and leaching losses with conservation tillage. Water quality is such an insidious and serious problem that a Presidential Water Quality Initiative was started in 1989 for the various agencies of the U.S. government to better understand the problem, and the implementation of solutions consistent with a prosperous agriculture.

ISSUES OF TECHNOLOGY AND RESOURCE CONSERVATION

Almost all of the technologies themselves have evolved in the 1940 to 1990 era primarily because of output vs. cost considerations, conservation benefits, environmental improvements, and lastly, their internal improvements. Earlier we mentioned incidental benefits and damages. As related to soil and water conservation, we discuss adopted technology issues of: tillage and residue management (consisting of crop residue return, conservation tillage, and soil quality response to tillage/residue management), crop rotation, machinery technology, weed management technology, crop improvement technology, fertilizer and crop nutrition management, and disease management technology. As these adopted technologies have changed, so have the uses of structures vs. management for soil and water conservation.

Tillage and Crop Residue Management

Conservation tillage is undoubtedly the main technology available and used for soil and water conservation in the United States, and in Canada as well. In fact, successes with tillage and crop residue management are actively being copied outside of North America (Carter, 1994; Jensen et al., 1994). Operationally, conservation tillage is any tillage-planting combination that retains at least 30% cover with surface residue after planting (Shertz, 1988). This definition is used by the CTIC (1991) for annual surveys of adoption. An abstract definition is a tillage system that maintains crop residue on or near the soil surface, a rough soil surface, or both, to control soil erosion and to provide good soil-water relations (Allmaras et al., 1991). Rationale for these definitions differs (Allmaras et al., 1991), but the common ground is that moldboard tillage is not a form of conservation tillage. The NRCS (CTIC, 1991) and ERS (ERS, 1994) use variations on the 30% cover definition in their adoption surveys, but the ERS survey also examines the form of primary tillage.

Current conservation tillage systems represent the merging of many necessary technologies, including the development of: (a) high producing crop cultivars, (b) machinery systems, especially those that facilitate tillage/planting into surface residues, (c) a set of principles required for operation of the tillage system to achieve soil and water conservation, and (d) pest (weeds, diseases, insects) control by means other than tillage alone. In this section, we will explain why the merging of these technologies has provided a system with major benefits to soil and water conservation, and has the potential to accelerate numerous benefits to production, the environment (including global carbon), and soil quality. Soil quality is considered by Larson and Pierce (1991) "as the capacity of a soil to function both within its ecosystem boundaries (e.g., soil map unit) and with the environment external to that ecosystem (extension into air and water quality)." An exhaustive discussion about soil quality (National Research Council, 1993) recognized the crucial input/control that tillage and crop residue management has over soil quality.

Crop Residue Return

There has been a steady improvement in production intensity since 1940 (Table 1); the primary intent was to provide more food and to sustain agricultural production and solvency. However, the potential impact on soil organic matter changes has not received public attention, rather attention is focused on losses of organic matter associated with grassland or forest conversion to arable agriculture—see Larson and Pierce (1991) for discussion about this issue and how arable agriculture must view the problem. Thus, the potential for soil organic matter improvements due to increased production might be viewed as an incidental or collateral benefit.

Yields of grain in the United States for seven crops have increased markedly in the 1940 to 1990 period (Table 2). Corn grain yields have made the largest increase (440% increase), whereas oat has shown the smallest change (75% increase). The harvest index (Table 2) is used to link grain yield to the total aboveground biomass, or biological production by the relation (Donald and Hamblin, 1976):

$$HI = Y_{gr}/Y_{biol} \tag{1}$$

where Y_{gr} is grain (or harvested) yield, and Y_{biol} is the total aboveground biomass including the grain.

Harvest index was originally intended as a criterion to compare among cultivars in plant breeding (Donald and Hamblin, 1976). Y_{biol} is usually estimated at grain harvest. Because leaf fall may occur before harvest, HI is a conservative maximum or overestimate of the crop-residue component of Y_{biol} in Table 2. Prihar and Stewart (1990, 1991) have shown that HI reaches a maximum when there is little or no plant stress associated with water and nutrient limitations. In most instances of plant stress, grain yield is reduced more than the vegetative biomass, and therefore HI in Table 2 may have a positive bias when estimating crop residue nationally. Prihar and Stewart (1990) estimated an HI of 0.48 to 0 53 and 0.58 to 0.60 for irrigated sorghum

Table 2. Grain Yield, Harvest Index, and Crop Residue Production for Seven Selected Crops over the 1940 to 1990 Period When Averaged for United States

	Grain Yield		Harvest Index		Crop Residue	
Crop	1940	1990	1940	1990	1940	1990
	kg/ha				kg/ha	
Barley	1280	2920	0.27	0.40	3460	4380
Corn	1890	7310	0.35	0.50	3510	7310
Oat	1150	2010	0.23	0.32	3850	4270
Sorghum	930	3790	0.34	0.50	1800	3790
Soybean	1260	2290	0.30	0.44	2940	2910
Sunflower	—[a]	1500	—	0.33	—	2760
Wheat	1050	2430	0.28	0.45	2700	2970

[a]No data reported.

and corn, respectively. For irrigated wheat they derived an HI range of 0.37 to 0.47. Consequently, the estimated crop residue (Table 2) is conservative.

Harvest indices (Table 2) were derived from reports made in the 1970 to 1990 era; an HI was not measured and reported in the 1940 era. Comparative HI in cultivars for 1940 and 1980 eras were measured under the same environment (Vogel et al., 1963; Hanway and Weber, 1971; Singh and Stoskopf, 1971; Donald and Hamblin, 1976; Buzzell and Buttery, 1977; Austin et al., 1980; Allan, 1983; Wych and Rasmusson, 1983; Wych and Stuthman, 1983; and Howell, 1990). These reports also provide tests/comparisons for evaluating growth and physiological contributions through genetic research. Harvest indices were increased about 45% with only minor differences among crops.

Potential crop-residue return (Y_r) was estimated (Table 2) by rearranging Eq. 1:

$$Y_r = Y_{gr}[(1/HI)-1] \qquad (2)$$

These crop residue returns increased even though HI had increased about 45% over the 1940 to 1990 era. Corn residue had the largest increase (3800 kg/ha) while soybean showed no significant change. Most of these increases in crop residue due to advances in crop technology had occurred by 1980, but yield increases since 1980 (Table 1) without significant HI changes have continued the increase of crop residue returns.

The significance of changing harvest methods on crop residue available for soil conservation can be demonstrated for wheat and corn. The reduced crop residue return before the development of combine harvesting, when harvest required transport of wheat bundles away from the site of plant development, can be estimated from different cutting heights (Douglas et al., 1989). When a semidwarf wheat 100 cm tall was cut at 17 cm vs. the ordinary combine cut at 50 cm, the crop residue return was reduced from 10.8 to 2.2 t/ha—a reduction of 80%. The actual crop residue return in 1940 was 540 kg/ha and would now be 590 instead of 2970 kg/ha, if the old harvest system were used. Components of dry matter in mature corn (Ritchie and Hanway, 1982) estimate that the corn picker harvest of corn cobs and husks along with the grain reduced field residue return by 33%; the real residue return in 1940 was 2340 kg/ha and would now be 4870 instead of 7310 kg/ha if corn pickers were used instead of a combine that removes only grain from the field.

Carbon return is crucial for long-term soil management/conservation. The amount of carbon available (C_r) in crop residues (Table 3) is estimated from data in Tables 1 and 2:

$$\text{C returned} = \text{C in shoot} + \text{C in root}$$
$$C_r = Y_{biol}(1 - HI)(0.4) + Y_{biol}(0.4)(0.2) \qquad (3)$$

It is assumed that most plant biomass other than grain is 40% carbon (Bidwell, 1974) and that root tissue (including rhizodeposition) produced during

Table 3. Estimated Changes (1940 to 1990) in Carbon Available in Crop Residue for Return to the Soil

	Shoot Only[a]		Root Plus Shoot[b]	
	1940	1990	1940	1990
Crop	C, kg/ha			
Barley	1380	1750	1760	2340
Corn	1400	2920	1840	4090
Oat	1540	1710	1940	2210
Sorghum	720	1520	940	2120
Soybean	1180	1160	1510	1580
Sunflower	—	1100	—	1430
Wheat	1080	1190	1380	1620

[a]Estimated 40% C content of crop residue.
[b]Root biomass estimated to be 20% of shoot biomass including harvested grain.

the vegetative and reproductive phases is 20% of the biological yield (Beauchamp and Voroney, 1994). This estimate is conservatively low. Buyanovsky and Wagner (1986) observed that root biomass was 70 to 80% of crop residues (wheat, soybean, corn) and contained 30% C. Their root C was therefore 25 to 30% of that in the biological yield. The "shoot only" carbon (Table 3) shows 50-year increases ranging from 108% for corn to no change for soybean; for "shoot-plus-root" carbon the increases range from 120% to 5%. There may be large variations of carbon return in crop residue depending on climate, soil productivity, and disease. For instance, Buyanovsky and Wagner (1986) reported returns of 3700, 3400, and 9200 kg/ha from wheat, soybean, and corn, respectively, in Missouri.

Field experiments mostly with moldboard tillage estimate that 15 to 22% of the carbon incorporated in the crop residue remains in the soil over the long-term (Larson et al., 1972; Rasmussen et al., 1980). A long-term perspective is that 10^4 kg C/ha in the upper 30 cm is needed to change the C content by 1%; to bring about such a change requires 15 years with corn and 35 years with soybean (Table 3).

Conservation Tillage

Current conservation tillage-planting systems were a long time under development (see Blevins et al., 1998). Perhaps the most significant machinery change required to adopt conservation tillage was the decision not to use the moldboard plow for primary tillage (Allmaras et al., 1985, 1991). Leighty (1938) made brief reference to use of a disk after oat harvest before planting corn in the subhumid and humid Corn Belt, otherwise it was moldboard plowing. Leighty (1938) mentioned stubble mulch in the semiarid wheat production. A formal stubble mulch tillage experiment was initiated cooperatively with Canadian scientists in 1941 (McCalla and Army, 1961). So there must have been adoption of nonmoldboard tillage as early as 1950 in the semiarid and arid Great Plains. Other research efforts to develop plow-plant in New York and no-till in Virginia, and the difficulty handling surface crop residue in the Southeast—all in the 1950 to 1960 era—indicated dominance of moldboard plowing (Allmaras et al., 1991). By 1980 (USDA, 1982) the USDA policy concluded that conservation tillage is the "leading cost-effective practice" for soil erosion; this projection was consistent with provisions of the Soil and Water Conservation Act of 1977. It was estimated that 15 to 25% of U.S. cropland was farmed with conservation tillage (USDA, 1982; Christensen and Magleby, 1983), and the definition was primarily centered on use of the moldboard plow (conventional) or another implement (conservation) for primary tillage.

A comprehensive tabulation of tillage adoption for 1993 (ERS, 1994) defines five tillage planting systems (Tables 4,5) used for wheat and soybean/corn production, respectively. The conv. w/mbd and conv. w/o mbd categories do not fulfill the definition of conservation tillage because they do not have 30% surface cover with crop residue after planting. These two categories distinguish use of a moldboard plow versus other tillage tools (disk, chisel, sweep) for primary tillage. For crop residue placement, this distinction is criti-

Table 4. Adoption of Tillage-Planting Systems for 1993 Planting of Wheat

Tillage System[d]	Winter Wheat[b]		Spring Wheat and Durum[c]	
	Planted Wheat[e]	Surface Cover	Planted Wheat[e]	Surface Cover
	%[f]			
Conv. w/mbd	6 (0-36)	2 (1-2)	8 (3-35)	3 (2-3)
Conv. w/o mbd	76 (44-85)	13 (9-18)	56 (43-78)	16 (12-18)
Mulch-till	14 (4-25)	39 (35-45)	27 (17-36)	42 (35-46)
No-till	4 (0-28)	54 (35-72)	7 (3-21)	61 (61-72)

[a]Adapted from ERS, 1994.

[b]States included are: CO, ID, IL, KS, MO, MT, NE, OH, OK, OR, SD, TX, WA.

[c]States included are: MN, MT, ND, SD.

[d]Conv. w/mbd uses the moldboard plow for primary tillage; conv. w/o mbd uses a disk, chisel, or sweep for primary tillage; conv. w/mbd and conv. w/o mbd have less than 30% surface cover after planting; mulch-till has a full-width tillage between harvest and planting; no-till has no tillage before planting; no-till and mulch-till have 30% or greater surface cover after planting.

[e]Percentage of planted wheat type.

[f]Mean (range); differences among states generates the range shown in parentheses.

Table 5. Adoption of Tillage-Planting Systems for 1993 Plantings of Soybean and Corn[a]

Tillage System[d]	Soybean[b] Planted Soybean[e]	Soybean[b] Surface Cover	Corn[c] Planted Corn[e]	Corn[c] Surface Cover
		%[f]		
Conv. w/mbd	8 (0-25)	3 (2-3)	9 (1-39)	2 (1-3)
Conv. w/o mbd	44 (34-82)	16 (8-20)	49 (35-60)	17 (14-19)
Mulch-till	25 (7-44)	40 (38-43)	24 (12-34)	38 (37-41)
Ridge-till	1 (0-3)	56 (56)	3 (0-17)	51 (34-53)
No-till	22 (6-38)	71 (69-75)	15 (2-27)	66 (59-76)

[a]Adapted from ERS, 1994.
[b] States included are: AR, IL, IN, IA, MN, MO, NE, OH.
[c]States included are: IL, IN, IA, MI, MN, MO, NE, OH, SD, WI.
[d]Conv. w/mbd uses the moldboard plow for primary tillage; conv. w/o mbd uses a disk, chisel, or sweep for primary tillage; conv. w/ mbd and conv. w/o mbd have less than 30% surface cover after planting; mulch-till has a full-width tillage between harvest and planting; no-till has no tillage before planting; no-till and mulch-till have 30% or greater cover after planting.
[e]Percentage of planted crop.
[f]Mean (range); differences among states generates the range shown in parentheses.

cal because the moldboard plow buries residue below 15 cm while the other tillage tools always maintain the residue above 10 cm (Staricka et al., 1991, 1992). Mulch-till, no-till, and ridge-till are conservation tillage-planting systems because they fulfill the 30% surface residue cover requirement—and they maintain crop residue in the upper 10 cm.

The Crop Residue Management System used in NRCS surveys is also reviewed (ERS, 1994); it has five tillage-planting systems. Three tillage planting systems (ridge-till, no-till, and mulch-till) are the same in both surveys, but the 15 to 30% and <15% surface-cover types do not key on primary tillage. Hence, the use of a moldboard plow versus other tillage tools for primary tillage is not identified when surface residue cover is <30% at planting in the Crop Residue Management System.

Tabulations in Tables 4 and 5 show a dramatic change in tillage systems away from the moldboard plow for primary tillage. Conservation tillage (the 30% cover definition) was used for tillage preparation and planting on 18, 34, 48, and 42% of the winter wheat, spring wheats, soybean, and corn, respectively (Tables 4 and 5). Moldboard plowing was used for 6, 8, 8, and 9% of the respective productions; the most used single category was conv. w/o mbd, which was used on 76, 56, 44, and 49% of the planted area. No-till in soybean and corn has captured 22 and 15% of the respective planted areas, but no-till in wheat is much lower. The ranges show the considerable variation of adoption among states. Mean surface-cover estimates for the different tillage systems are characteristic—the ranges reflect variations in residue from the previous crop and amounts of secondary tillage, especially for the mulch-till and conv. w/o mbd systems.

One of the most important summations of planted area is the combination of conv. w/o mbd, mulch-till, ridge-till, and no-till for winter wheat, spring wheats, soybean, and corn—these are 94, 92, 92, and 91%, respectively. These percentages are the ones in which the crop residue is either on the surface or buried at a depth less than 10 cm. Although the adoptions in Tables 4 and 5 do not indicate tillage rotation, they suggest that the conv. w/ mbd is not rotated frequently with the other four systems. This adoption trend may have great significance for soil quality if it can be shown that the crop residue decomposition is changed by the retention of residue at or near the soil surface.

Soil Quality Response to Tillage/Residue Management

There are both short-term and long-term impacts of crop residue return (Table 3) and the tillage system (Tables 4 and 5) on soil quality. These impacts can be observed in experiments when the same form of primary tillage, or no-tillage, is used continuously. A system without tillage retains crop residue mostly on the surface and some within the upper 5 cm due to faunal activity and the exponentially declining root residue with depth; systems that use a disk, chisel, or sweep for primary tillage mix most of the new with the old crop residue in the upper 10 cm and leave some on the surface; and a moldboard plow buries the most recent crop residue below 15 cm while returning last year's crop residue and some root tissue to the top 10 cm (Staricka et al., 1991, 1992; Allmaras et al., 1996a).

Tillage systems were shown to have significant short-term impacts on soil organic carbon and water stable aggregation in a cool-humid climate for barley production (Angers et al., 1993a and b); tillage systems compared were no-till (NT), chisel plow (CP), and moldboard plow (MP). After 4 years of test, C contents in the upper 15 cm were greatest in the CP and least in the NT and MP tillages, but the contents of microbial biomass C, acid-hydrolyzable carbohydrate, water soluble carbohydrates, and alkaline phosphatase all followed the sequence NT>CP>>MB (Angers et al., 1993a). Water stable aggregation in the 0 to 7.5-cm layer was in the order NT>>CP>MB (Angers et al., 1993b) and correlated positively with microbial biomass C and water soluble carbohydrates, both of which may influence infiltration. A more comprehensive analysis of the short-term influence of tillage systems on organic carbon reactions/composition is given by Angers and Carter (1995). These differences in soil organic matter produced over such a short time by tillage systems in a cool-humid climate require a longer time period in the warmer and subhumid semiarid climates where decomposition rates are greater, there may be periods when readily decomposable organic matter is depleted, and the organic carbon compounds are less labile (Angers and Carter, 1995).

There are numerous observations linking increased water stable aggregation as related to polysaccharides produced during decomposition of crop residues (Allmaras et al., 1996b). When these polysaccharides and the increased water stable aggregation are located near the soil surface, infiltration has been increased (Bruce et al., 1992). So the marked change to tillage systems that retain crop residues in the upper 10 cm (Tables 4 and 5) has had a marked short-term change in soil water relations especially in the subhumid to semiarid regions. The magnitude of this change of infiltration and the benefit of water conservation and use has undoubtedly influenced the adoption of conservation tillage, but there are no current industry estimates (Allmaras et al., 1996b).

Another short-term impact on soil quality is the change in wind/water erosion potential due to the tillage system itself and how it disposes of the residue. Components of the C factor (cropping factor) in USLE and RUSLE are the amount of surface cover, the surface random roughness, and the amount of crop residue present in the 0 to 10-cm layer (Laflen et al., 1985)—these are all estimable as related to tillage systems (Allmaras et al., 1991). However, the surface roughness and shallow buried crop residue components may be changed when crop residue is always retained on the surface or buried at a shallow depth. For more detail on water erosion predictions with RUSLE, see Renard et al. (1991). The prediction of tillage system and crop residue management impacts on wind erosion were developed using concepts of ridge (oriented roughness), standing residue, random surface roughness equivalent to nonerodible aggregates with diameters >0.84 mm (Skidmore, 1994), and has more recently been improved to utilize tillage/residue management controls over soil structure, soil crusting, and vegetative cover to reduce wind drag on the surface (Skidmore et al., 1994).

This estimated short-term influence of conservation tillage has already occurred in the period from 1982 to 1992 (Kellogg et al., 1994). The predominance of moldboard plow primary tillage in 1980 has given way to nearly 95% dominance of nonmoldboard systems (Tables 4 and 5). During the 1982 to 1992 period, sheet and rill erosion in cropland has fallen from 9.2 to 6.9 t/ha per year—a reduction of 25%. Over the same period, wind erosion has fallen from 7.4 to 5.4 t/ha per year—a 27% reduction. These are the largest 10-year estimated changes since NRI records were started about 1965.

Long-term benefits (nominally more than four years depending on hydrothermal climate) are somewhat preliminary because analytical procedures for carbon including the natural isotopes are new; field soil sampling techniques are also much improved. Reicosky et al. (1995) summarized rates of soil organic matter changes (kg/ha per yr). There were moderate crop rotation effects but the no-till vs. moldboard tillage comparison was outstanding. Rates for no-till could be somewhat positively biased because the tests were not started when there was a high residual organic matter. The negative rates with moldboard tillage were nearly zero only when large amounts of residue were returned. Karlen et al. (1994) and Lal et al. (1994) found greater soil carbon contents in no-till and chisel-based compared to moldboard-based systems in the Ap layer. The length of tillage treatment was about 10 years, and both locations have a subhumid to humid temperate climate.

Lamb et al. (1985) showed that 12 years of wheat fallow had decreased soil carbon by 4, 14, and 16% for no-till, stubble mulch, and conventional tillage, respectively, when carbon in native grassland is used as the base. Havlin et al. (1990) showed larger C contents in the 0 to 30-cm layer after 11 years of no-till compared to a chisel-disk based system; the difference in the two soils was greater in continuous sorghum than continuous soybean—a residue return effect somewhat related to Table 3.

The rate of organic matter change for the moldboard vs. sweep based systems in a wheat-fallow rotation of Pendleton, OR is shown in Table 6. With moldboard primary tillage, the C change rate was negative until 20 t/ha of manure was applied for each wheat crop. The negative C change was reduced by 50 kg/ha per yr when a sweep was used instead of a moldboard plow and 45 kg N/ha was applied; all other tillage operations were the same. When the N application rate was raised from 45 to 180 kg/ha using moldboard tillage there was a positive C change of 70 kg/ha per yr, but when a sweep was used the change was an additional positive 100 kg/ha per yr. There are differences in

Table 6. Long-Term Rates of Carbon Change as Related to Primary Tillage and Residue/Amendment Additions (1931–1986, Pendleton, OR)

Primary Tillage[a]	Fertilizer/Amendment Management[b]	Carbon-Change Rate[c] (kg/ha per yr)
Moldboard	no N added	-190
	no N (fall burn)	-210
	45 kg N/ha	-130
	2 t/ha pea vines	-100
	20 t/ha manure	+30
Sweep	45 kg N/ha	(-80)
	180 kg N/ha	(+170)
Moldboard	180 kg N/ha	(+70)

[a]Primary tillage preceding summer fallow; moldboard tillage depth = 18 to 24 cm; sweep tillage depth = 10 cm; secondary tillage same for all treatments and always less than 10 cm.

[b]Amounts shown were applied per wheat crop alternated with fallow.

[c]Estimates without parentheses made from C measurements in 0 to 30-cm depth from 1931 to 1986 (Rasmussen and Parton, 1994); estimates in parenthesis from C profiles in an adjacent field study (Allmaras et al., 1988); and the 45 kg N with moldboard plow were matched for comparison between studies.

residue production and return due to N application, but residue production responses to tillage are much smaller (Rasmussen and Parton, 1994).

These long-term carbon changes demonstrate that the recent changes of tillage away from conv. w/mbd (Tables 4 and 5) could have already modified C storage in the managed biosphere since the 1980 era, as discussed recently for the whole biosphere (Ciais et al., 1995; Keeling et al. 1995). Because nutrient adjustments can be made in managed ecosystems, past and future plant-biomass increases in response to elevated CO_2 are more likely than in natural ecosystems (Culotta, 1995). Root and rhizosphere responses to elevated CO_2 are likely to be greater than shoot responses (Rogers et al., 1994).

Comparative Adoption of Soil Conservation Practices

Conservation tillage was declared to be a soil conservation practice in about 1980 (Allmaras et al., 1985). Within a span of only 10 years, it has become the single most acclaimed practice for soil conservation (mainly wind and water erosion control). About 23% of the 155 million ha of cropland, excluding the Conservation Reserve Program (CRP), had conservation tillage (30% cover rule) in 1992 (Magleby et al., 1995). Other soil conservation practices applied on cropland were: contour farming, 8%; field windbreaks, 2%; grassed waterway and outlets, 6%; terraces, 9%; strip cropping and contouring, 1%; and pasture/hayland management, 17%. Only conservation tillage, grassed waterways and outlets, and pasture/hayland management showed significant growth since 1982 (Magleby et al., 1995). Contour farming actually decreased. Any given parcel of land could have had more than one practice. When conservation benefits are assigned to crop residue on or near the surface as in wind and water erosion research, all four classes of tillage other than the conv. w/mbd in Tables 4 and 5 may be given conservation tillage status. This raises to 70% of cropland protected by conservation tillage. In the compilation there were 13.8 million ha of CRP land with grass, tree, or wildlife habitat cover (Magleby et al., 1995).

Studies in Ohio show more adoption of soil conservation practices on larger farms based upon gross sales (Stout et al., 1989). Yet, conventional tillage in 1986 remained at 65% for the smallest size farm and had been reduced to 35% on the two largest of the seven farm sizes. As farm size increased there was a strong increase in percentage of farms using two or more of soil conservation practices including contour plowing, terracing, strip cropping, cover crops, and grassed waterways/diversions. This use of soil conservation practices on the larger farms is even more remarkable because the smallest farms had a larger share of steeper sloping lands in their farm unit, although forage production had a higher share of their cropland. Stout et al. (1989) discuss lower technical efficiencies and greater financial stresses on the smaller farm units.

Pierce et al. (1986) examined the C and P factors and the RKLS factor combination of the Universal Soil Loss Equation (USLE) as recorded in the National Resources Inventory for 1982. The C factor relates to soil roughness and soil cover as in conventional vs. conservation tillage, the P factor relates to supporting practices such as contour operations, or divided slope, and RKLS (t/ha per year) is a combination of factors that determines natural or potential erosion. Terracing also reduces erosion because it is a component of the P factor and reduces the LS (slope length and steepness) factor. Interestingly they showed a need for using the C, P, and LS factors on some lands; i.e., more soil conservation practices than conservation tillage alone on lands with higher RKLS. They also noted that conservation tillage (low C value) was being used for more than erosion control in lands with low RKLS; was there some improved infiltration and water storage? Was mere economic gain involved?

Crop Rotation

Crop rotation is a time sequence of different crops on the same parcel of land; the length of a rotation is the number of full-season crops in a cycle of the time sequence. When cover crops and double cropping are used in climates where the cold season permits plant growth, the rotation length includes cover crops, double crops, and full-season crops in a cycle. Karlen et al. (1994) and Reeves (1994) present extensive reviews of crop rotation history and function. Lengths of crop rotations are often referred to as monoculture (no rotation/diversity), short rotations (e.g., corn and soybean alternating), or extended rotations (e.g., corn, oat, meadow/clover/alfalfa).

Technology adoption, including conservation tillage and associated crop residue management, has changed crop rotations and their need, function, and economic return. Crop rotation with sod and cover crops has always had direct influences on soil erosion and physical properties of the soil (Reeves, 1994). Without herbicide assist, tillage to prepare a seedbed for an alternating cash crop after a sod crop was excessive enough to destroy some of the soil structure improvements provided by the sod crop. Extended rotations with legume and forage crops (corn, oat, clover, meadow) were necessary before 1940 for N fertility and animal feed including draft horses. With the development and use of N fertilizers and tractors, crop rotations after 1940 were shortened, and there was a search for more cash crops. With advent of chemical pest control and the shortening of crop rotations, the age-old value and capability of crop rotation for cultural pest control has partially diminished.

Although there was some monocropping of wheat (wheat-fallow) and corn prior to 1940, monocropping was a highly developed technology practiced in the 1950 to 1960 era (Pearson, 1967; Power and Follett, 1987). A highly capitalized, mechanized system was developed around the management skills and workload to be handled by a single operator—such a system did not lend itself to the diversified practices needed for the mix of crops in a crop rotation.

Tweeten (1995) indicates that crop rotations on smaller farms in Ohio have more crops than in larger farms, based upon sales of commodities.

Short crop rotations without multiple-year forage or sod-forming crops dominates arable farmlands. The current practice of crop rotation was surveyed (ERS, 1994) at planting of corn, soybean, and winter/spring wheat (Tables 7 and 8) in 1993. Surveyed were the crops that preceded these four crops in 1991 and 1992. Percentages in parentheses (Tables 7 and 8) represent the range found among states included in the survey. These ranges of percentages related to crop rotations are much larger than those in a compilation of ranges to describe how the same states varied in their use of tillage systems (Tables 4 and 5).

As much as 82% of the planted corn/soybean in 1993 was either monocropped or alternated between these two row crops (Table 7). Only a very small percentage of soybean followed soybean, but about 25% of the corn followed corn in NE, WI, and MI, whereas as much as 22% of soybean followed soybean only in AR and MO. An unusually high 40% of corn in SD had small grain in 1992 and/or 1991. Only 5% of planted corn/soybean was planted where small grains were grown in 1991 and/or 1992. The only state with a significant rotation of hay with corn (37%) was WI. Nearly all of the double-cropped soybean was in AR and MO (Table 7).

Over 86% of the winter wheat was continuous or rotated with fallow (Table 8). Continuous winter wheat was concentrated in KS, OK, TX with a weighted mean of 62% of the winter wheat planted in 1993. Winter wheat in a rotation with fallow was greater than 70% of the planted winter wheat in each state of CO, ID, MT, NE, OR, SD, and WA—the weighted mean was 86%.

Table 7. Current Practice of Crop Rotation in a 3-Year Period Including the 1993 Planting of Corn and Soybean[a]

Three-Year Crop Sequence[b]	Corn		Soybean	
	(% of planted crop)[c]			
Continuous crop	25	(10–59)	6	(0–22)
Continous other row crops	58	(19–76)	76	(7–96)
Small grains with row crops	5	(0– 40)	5	(0–32)
Idle in rotation	8	(2–28)	6	(1–22)
Hay or other crops	5	(2–37)	1	(0–3)
Double-cropped soybean	—	—	6	(0–37)

[a]Adapted from ERS, 1994.
[b]The three-year sequence included the 1991, 1992, and planted 1993 crop.
[c]Percent of 1993 planted crop; recorded mean is weighted by planted land in each state selected; numbers in parentheses are the range of percent for states included in the survey.
corn: IL, IN, IA, MI, MN, MO, NE, OH, SD, WI;
soybean: AR, IL, IN, IA, MN, MO, NE, OH.

Table 8. Current Practice of Crop Rotation in a 3-Year Period Including the 1993 Planting of Winter Wheat and Spring Wheat[a]

Three-Year Crop Sequence[b]	Corn		Soybean	
	(% of planted crop)[c]			
Continuous wheat	38	(0–94)	14	(3–17)
Continuous other small grains	<1	—[d]	12	(1–16)
Row crops with small grains	15	(0–92)	36	(0–74)
Fallow in rotation	48	(2–98)	35	(12–84)
Hay or other crops	<1	(0–8)	3	(0–5)

[a]Adapted from ERS, 1994.
[b]Three-year sequence: 1991, 1992, 1993 planted crop.
[c]Percent of 1993 planted crops; recorded mean is weighted by planted land in state selected; numbers in parentheses are the range of percent for states included in the survey.
winter wheat: CO, ID, IL, KS, MO, MT, NE, OH, OK, OR, SD, TX, WA;
spring wheat: MN, MT, ND, SD.
[d]No range reported.

Only in IL, MO, and OH was more than 70% of the winter wheat seeded into a rotation of small grain and row crops—the weighted mean was 83%. Practically no winter wheat is grown in rotations with other small grains. In contrast to winter wheat, only about 14% of spring wheat was continuous, and only 35% was rotated with fallow (Table 8). In the more humid part of the spring wheat region (SD and MN) there is as much as 60% of spring wheat planted into rotations with row crops—the mean is 36%.

Crop rotation is highly regionalized whereas the adoption of tillage systems is not significantly regionalized. Apparently the nonmoldboard systems are being used successfully in all crop rotations. The scale of regions was large in the data bank (Tables 4, 5, and 7, 8), but a similar conclusion was evident when evaluating at a smaller scale of adoption using other crops with the dominant corn, soybean and wheat crops (Allmaras et al., 1994). The dominance of alternating soybean and corn, alternating wheat and fallow, and even the significant amount of continuous corn/wheat all at the expense of longer rotations indicates that fertilizers and chemicals for pest control have replaced some of the older (pre-1950) needs for crop rotation. Many benefits from crop rotation can still be cited (Karlen et al., 1994; Reeves, 1994), but there are many economic, managerial, and governmental policy factors that encourage monoculture, especially the wheat and fallow type. Crop rotation is perhaps the most notable technology wherein practice and research recommendation differ.

The "rotation" effect is not observed when the component crops are grown in monoculture (Crookston et al., 1991). Crookston et al. (1991) demonstrated first-year corn/soybean (after 5 years of the monoculture) yielded at least 15%

more than their monoculture, and at least 8% more than alternating corn-soybean in two fine-textured soils. A moldboard-plow based system with fertilizer and herbicide were used to manage at a high but not excessive level. Numerous follow-up measurements (Crookston, 1996) identified related growth, function, and environment of the roots, yet the "rotation" effect was not fully clarified. Enough information was obtained to encourage a longer rotation; presumably an annual cereal or legume as a third crop can improve root ecology in this moldboard-based system. Meese et al. (1991) have observed similar first-year and alternating soybean-corn effects in a moldboard-based system, but additional years of continuous soybean were more deleterious because of brown stem rot (*Phialophora gregata*) in a susceptible cultivar. Adee et al. (1994) observed more disease and yield decline in soybean in the short rotations with a moldboard-based system than observed by Crookston et al. (1991); the decline was more serious in the no-till system and was linked to disease inoculum carryover, yet there was a "rotation" effect similar to that observed by Crookston et al. (1991). The Wisconsin team also concluded that soybean should be separated by two instead of one year. Undoubtedly these studies would have had more difficulty demonstrating a rotation effect had there been a major difference in crop water utilization and failure to recharge the root zone during the cold season as in semiarid areas.

A benefit of crop rotation is the control of pests while continuing with conservation tillage innovations and reducing chemical use (Karlen et al., 1994; Reeves, 1994). Guidelines for evaluating the potential of crop rotation to control pests were clearly detailed by Flint and Roberts (1988) as feasible only for certain types of pests and situations related to ecology of the pest and agronomic/economic situations. The source of the pest must be from within the field and cannot move from areas adjacent to the field; the host range of the pest must not be so wide as to exclude practical alternate crops; and the pest population cannot remain in the absence of a living host. All three requirements must be met and within economic/managerial reality. Flint and Roberts (1988) discuss several crop rotations in California agriculture mandated in the absence of other control mechanisms, such as chemicals or a pest resistant cultivar.

Machinery or Mechanical Technology

Agricultural machinery development was already occurring in the nineteenth century (Cochrane, 1993) before there were significant biological and chemical technologies to be adopted; continued development and adoption of machinery technology has facilitated the adoption of these other technologies. Without major machinery technologies, conservation tillage and farming systems could not have become the centerpiece of soil and water conservation. These machinery advances cover a broad spectrum of materials, methods, and equipment. This section will focus on advances in farm machinery and their role in adoption of soil conservation practices in the last 60 years.

Since 1940, agricultural machinery systems have moved from relatively small labor-intensive systems to large complex systems requiring a high degree of knowledge and skill to operate, maintain, and repair. Equipment became larger and capable of functioning at higher ground speeds in the 1950 to 1970 era. High-priced manual labor was replaced with machines that increased individual farmer productivity and operational timeliness needed for economic survival.

The average power of tractors sold in the U.S. increased dramatically from 20 kW in 1950 to over 50 kW in 1982 (USDA, 1994). Demands for increased tractor power have diminished; no further increase in tractor power is projected for the year 2000 (Hood et al., 1991). These larger power units pull larger tillage, planting, transporting, and harvesting machines often at faster speeds as needed to increase timeliness and expanse of land per operator. Following are some of the milestones in tractor development: 1940s—engine driven power takeoff, self starters, and hydraulic systems; 1950s—power steering, and power shift transmission; 1960s—closed-center central hydraulic system, hydrostatic transmission, and articulated with four wheel drive; 1970s—turbo charged diesel engines, and front-wheel assisted drive; and 1980s—low pressure radial tires, onboard computer systems, and rubber tracks (Goering, 1989). These technological changes delivered engine power to the appropriate mechanical operation; they also improved operator safety, operator comfort to reduce fatigue and extend the workday, and implement performance.

These technological advances in tractor design have impacted adoption of conservation practices positively and negatively. Larger power units provided more timely tillage, planting, and harvesting so critical for development of conservation tillage systems. Options for pest control with tillage, fertilization, and seedbed preparation are restrictive in no-till and other nonmoldboard tillage systems. Furthermore, these operations are often in a one-pass operation. Improved tractor hydraulic systems and computer monitoring of speeds, wheel slip, and draft, now allow more accurate implement control on soil tilth, seed depth, soil roughness, and other environments for soil conservation. On the negative side, tractor size has increased the potential for soil compaction that often restricts root development (Voorhees, 1992), limits water infiltration (Pikul and Allmaras, 1986), and increases incidence of soilborne plant disease (Allmaras et al., 1988). Axle weight, contact pressure, and soil conditions are major factors controlling the degree of soil compaction (Voorhees, 1992). Axle weight of tractors has increased as much as 500% with tractor power in the last 60 years but contact pressure of wheel tractors has not changed greatly and so subsoil compaction has increased. This increased axle weight is often problematic in spring when soil is wet and more conducive to compaction. Advances in low pressure radial tires have improved traction and reduced tire-soil contact pressure; both can reduce soil compaction (Raper et al., 1995).

Harvesting equipment has changed in size and complexity similar to technological advances in tractors. For example, corn was typically harvested with a tractor mounted two-row corn picker in 1945 but today self propelled combines are used with as many as eight-row corn heads. A similar situation in wheat harvest has occurred, in which binders and header harvesters have been replaced with combines. Cereal grain combines have also increased in size so that 10-m wide headers are not uncommon. These large and heavy combines with large axle weights present soil compaction problems that may be worse than with large tractors. This is especially troublesome when soil is wet and optimum for compaction during harvest (Clark and Langdale, 1990).

Several major changes in combine technology during the last 60 years were the introduction of axial flow cylinders and stripper-headers. Both have increased machine capacity and therefore more harvested land each day per farm operator. Stripper-headers strip grain from cereal crops with comb-like teeth mounted on a rotating drum—they leave stems remaining upright and anchored to the soil. Stripper harvesting has potential soil and water conservation benefits (Wilkins et al., 1996) because the intact cereal crop stems trap more snow, reduce wind velocity at the soil surface, and provide shading needed to reduce evaporation in the cool season. Another significant contribution to adoption of conservation farming systems was the improved straw and chaff spreading devices attached to grain combines (Douglas et al., 1989). Uniform crop residue distribution is essential in no-till and other non-moldboard systems. Excessive straw and chaff left in rows by combines can easily increase the residue concentration 3 fold and thus influence nutrient uptake and supply, interfere with weed control and seeding, and harbor plant disease inoculum (Douglas, et al., 1992).

Advances in tillage and planting equipment during the past 60 years have directly impacted management of crop residue for soil erosion control and other benefits to soil quality. Crop residue may provide excellent erosion control, both when on the surface or buried near the surface, but it had presented significant challenges for machinery system design. The first steel moldboard plow developed by John Deere in 1837 was an improvement over cast-iron plows that failed to invert the furrow slice and bury plant residue in sticky prairie soils. Moldboard plows served as the primary tillage in most farming systems for more than 125 years. Their popularity grew as crop production increased, until the peak production of moldboard plows in the U.S. occurred in the 1950s and 60s when 75,000 to 140,000 units were shipped annually (USDA, 1965; USDA, 1977). Gradually the moldboard plow has been replaced with chisel plows, sweeps, and disks for primary tillage; in some cases primary tillage has been eliminated. The ratio of moldboard to chisel plows shipped in the U.S. has decreased from 31.5 in 1955 to 0.4 in 1990. In the late 1980 through early 1990 period, fewer than 3,000 moldboard plows were shipped annually in the U.S. Many studies have shown that the moldboard plow buries weed seeds and crop residue, creates macropores/random roughness, and provides soil conditions optimal for secondary tillage and planting with equipment not designed to clear crop residue. Although increased macropores and random roughness are desirable for soil erosion control, a soil surface devoid of crop residue is not desired for soil erosion control and soil organic carbon accumulation. Moldboard plowing places crop residue below 10 cm compared to chisel plowing that leaves residue in the top 10 cm (Wilkins and Kraft, 1988). The long-term benefits of keeping crop residue near the soil surface by conservation tillage with sweeps, disks, or chisels for primary tillage has a major positive impact on soil carbon storage (Table 6). Farmer and public demand to improve soil erosion control in the 1970 through 1980 period prompted a major replacement of the moldboard plow with tillage systems that leave crop residue on and near the soil surface. The Food Security Acts of 1985 and 1990 required producers with highly erodible crop land (HEL) to follow farm conservation plans for minimizing soil erosion and maintaining eligibility to participate in USDA programs. Most of these farm plans required 30% soil surface covered with crop residue at planting for erosion control.

The combination of farmer stewardship and the Food Security Acts has precipitated major changes in tillage and planting systems. Drastic modifications and alterations to clean-till equipment were made since the 1970s for maintaining residue on the surface in conservation tillage systems. Crop residue wraps around rotating components, wedges into openings, collects where clearance is limiting, hairpins in seed furrows (Erbach et al., 1983), especially when the residue is wet (Choi and Erbach, 1986). Vertical clearance as well as lateral and longitudinal spacing between soil engaging tools has increased. Coulters of various shapes and adjustments have been added to cut residue ahead of shanks. The number of field passes has been reduced to save energy and retain surface residue. Two or more field operations have been combined into a single pass. Examples are cultivating and applying fertilizer, chopping residue and tilling, tilling and seeding, or fertilizing and seeding. Air seeders were originally designed to combine tilling and seeding and now there are options to combine tilling, fertilizing, and seeding. The chisel-planter combines tilling, seeding and fertilizing into a one-pass operation (Peterson et al., 1983)

Maintaining crop residue on the surface for erosion control has required significant modification of seeding equipment. The two most obvious changes are increased mass and clearance. Seeding equipment requires more downward force in no-till and other forms of reduced tillage because of soil strength and more soil engaging tools per seeded row. The seed zone is drier and harder during fall seeding in chemical-fallow fields in the PNW as compared to clean tillage of summer fallow fields (Hammel et al., 1981; McClellan, 1987), but soil wetness and soil consistence (plastic range) can be problems in more humid climates. Devices have been added to seeders to move crop residue away from openers and the seed furrow. These include coulters to cut residue ahead of openers, and scuffer wheels or angled disks to move residue to the side (Payton et al., 1985; Hyde et al., 1987; Morrison et al., 1988). Triple disk openers, large diameter disks, offset double disks, and angled single disks have been incorporated to improve cutting through surface residue (Tessier et al., 1991). Some seeding equipment has been developed to meet special requirements of conservation tillage systems. Examples are the Yielder® drill that was developed to seed through hard dry conditions in the Pacific Northwest (Hyde et al., 1987), till-plant systems to plant in a ridge furrow system, cross-slot seeder to seed into sod and standing stubble (Baker and Saxton, 1988; Wilkins et al., 1992).

To improve fertilizer use and reduce field passes with tillage, seeding equipment has been modified to place fertilizer either below or below and to the side of the seed (Papendick et al., 1985, Hyde et al., 1987, Wilkins et al., 1987; Wilkins, 1988; Morrison and Potter, 1994). Various tool configurations have been developed for incorporating fertilizer with a minimum of soil disturbance and a maximum retention of crop residue on the soil surface (Morrison et al., 1988). Fertilizer placement is critical to avoid seedling toxicity and assure maximum fertilizer uptake. Fertilizer toxicity is related to source, application rate, soil water potential, and soil temperature (Mahler, 1985; Mahler et al., 1989). A minimum separation distance between seed and fertilizer is 5 cm (Babowicz et al., 1983). Wilkins and Haasch (1990) found stand reductions when winter wheat seed was separated 2.8 cm from an 80 kg/ha band of ammonium nitrate in a dry silt loam. Baker et al. (1989) developed a spoke-wheel fertilizer injector for conservation tillage systems. This method allows fertilizer to be placed into the soil with a minimum of soil and surface-residue disturbance. Womac and Tompkins (1990, 1991) developed a kinematic inverted slider crank mechanism to operate a single probe that entered and exited the soil without the disturbance associated with a spoke wheel.

A recent farm equipment development is automatic control to improve performance and provide variable rates of pesticide/fertilizer/seed within a field. Tractors have digital readouts on speed, fuel efficiency, and slip. Seeders, drills, and sprayers have monitors and warning devices that activate when equipment performance fails to meet preset standards. These improvements have emerged from *precision farming* technology which is using equipment performance to optimize production factors in all parts of the field. Global positioning systems (GPS) can generate precise yield maps (Macy, 1994) to improve fertilizer use (Peterson, 1991) and weed control (Rudolph, 1994). These technological advances have and will continue to improve conservation farming systems.

Weed Management and Control Technology

Research in the early 1940s on 2,4-D [(2,4-dichlorophenoxy) acetic acid] in the United States and on MCPA [(4-chloro-2-methylphenoxy)acetic acid] in England, their release after World War II, and their rapid acceptance by farmers gave worldwide attention to selective weed control. Phenoxy herbicides have provided economical and selective postemergence control of broadleaf weeds in Graminaceae crops and noncropland for the past 5 decades. This technology has had a major impact for increased crop yields.

Total numbers of basic herbicides in the United States were 14 in 1940, 25 in 1950, 61 in 1960, 131 in 1970, 156 in 1980, and 145 in 1990. Herbicide technology has switched from inorganic to organic herbicides, nonselective to selective herbicides, postemergence to soil-applied (some incorporated) herbicides and now back again, and single herbicides to mixtures and even multiple applications. Various additives are used to enhance herbicidal activity and to prevent abnormally early degradation. Selective herbicides are a major method of weed control. Mechanical weed control was being steadily replaced by herbicide technology and reduced tillage until about 1980, when environmental concerns reversed the trend somewhat toward a balance of chemical and mechanical weed control. These trends can be observed in recent tillage systems being utilized and the modest decline in amount of herbicide used (ERS, 1994) on a mass basis.

Technology for weed control is changing with increased emphasis on : (a) biocontrol of weeds, (b) improved ecological knowledge of weeds and their reproductive and disseminative strategies in cropping systems, (c) recently-developed herbicides at lower application rates, (d) more precise equipment for tillage and chemical application, (e) herbicide formulations with reduced persistence and toxicity in soil and aquatic environments, (f) variable rate application technology to use herbicides only where weed infestations warrant control in a field, and (g) herbicide tolerant crops to improve weed control and crop selectivity. Water quality (surface and groundwater), as well as safety of food, wildlife, and human health, are improved by these new and improved weed management technologies. A new paradigm has public and private sector scientists collaborating to provide alternative weed control technology and improved herbicide technology, respectively, for more balanced and integrated systems of weed management.

Management of Weeds in Conservation Tillage Systems

Weed control is often a major production expense for crop production with reduced tillage methods. A given land area has the capacity to produce a given amount of plant dry matter during the growing season. If 15% of the dry matter is weeds, then crop biomass will be reduced correspondingly because crops and weeds compete directly for water, nutrients, and light. Following is a brief outline of weed control principles developed through the 1940 to 1990 era of weed control (Wicks et al., 1994).

Classification of Weeds

Weeds may be classified based upon their life cycle as annuals (summer or winter), biennials, or perennials (simple, bulbous, or creeping). Weeds in conservation tillage typically include winter and spring annuals plus bulbous and creeping perennials. Perennial weeds often increase as tillage is reduced. Weeds may be further classified according to their response to tillage. *Arable response weeds* require periodic tillage of the soil for survival (e.g., many annuals), *inverse response weeds* require nondisturbed soils for survival (e.g., biennials and perennials), and *intermediate response weeds* survive in either tilled or untilled situations (e.g., creeping perennials). When there is tillage, weeds are either arable or intermediate response weeds, and when untilled, inverse and intermediate response weeds prevail.

Strategies for Weed Management

Weed control in conservation farming systems may control the choice of crops, crop rotations, date and rate of seeding, row spacing, method of cultivation, and herbicide formulation—special cultural weed control practices must often be used. Ridge tillage is a special conservation-tillage system that leaves plant residue and weed seeds on the soil surface but depends on both postemergence cultivation and herbicides banded only over the row (Forcella and Lindstrom, 1988). Adoption of this weed management strategy has reduced herbicide use (ERS, 1994).

Weed management strategies in conservation tillage systems may be to do nothing, practice economic weed control, or prevent weed seed production. High costs of control or minimal immediate risk to production may prevent control of weeds in roadsides or other areas and late-emerging weeds in conservation tillage systems. With economic weed control, financial returns must exceed the cost of control (Lybecker et al., 1991). Thus, a certain population of weeds may be tolerated if control is uneconomical. Others approach weed control by eliminating seed production, which involves greater vigilance and expense (King and Oliver, 1994). Once the weed management strategy is selected, it will dictate subsequent weed management objectives.

Weed-Management Objectives

The conservation tillage practitioner may prevent, tolerate, eradicate, or control weeds. Prevention involves the exclusion of all weeds or weed propagules from an area so that one is no longer confronted with that weed species. The adage 'an ounce of prevention is worth a pound of cure' is highly relevant for weed management. However, once the weed has invaded an area, one of the latter three weed management objectives is required. Tolerance means allowing certain weed species because they may be causing little or no yield loss, procedures for control are not economical, or there is not ample time for control. Weed eradication is the complete elimination of all live weeds or weed propagules from an area. Eradication is a popular objective in regulatory decisions, but it is nearly impossible to accomplish in field crop production. Limited success with eradication by soil fumigation has been used to control propagules of nutsedge and other weeds in high-value vegetable crops. The persistence of weed species and our limited ability to eliminate all weed propagules have prevented regulatory agencies from eradicating even a single weed species from a large area. Weed control is in reality a default process of limiting weed infestations so that crops can be grown profitably or other operations can be conducted efficiently. Most of what can be done to mitigate weeds falls under this weed management objective.

Weed Control Methods

Weed control methods for conservation tillage can be broadly classified as mechanical, cultural, biological, chemical, and integrated.

Physical or cultivation control of weeds still represents the major weed control method. Conservation tillage practices often have particular needs for tillage operations; e.g., row spacing may dictate method of cultivation. Physical control methods may remove weeds, bury them, or weaken them by partial root or foliage destruction. Hand pulling, spudding, or hoeing may control weeds better than mechanical tillage but cannot often be used because of timeliness and labor costs. Tillage implements, such as plows, disks, sweep plows, rotary cultivators, sweep cultivators, field cultivators, rotary hoes, harrows, and finger weeders, vary in the amount of weed residue to be removed and weed roots to be destroyed. Physical methods also include mulching, mowing, cutting, flooding, dredging, draining, chaining, and the use of heat, lasers, and ultrasonics.

Cultural weed control exploits plant competition because the first plants to occupy an area have a competitive advantage over latecomers. Thus, winter annual crops compete quite successfully with subsequently germinating summer annual weeds. A rapidly germinating and emerging crop cultivar gives a competitive edge. The ability of close-seeded soybean to compete is one factor prompting no-till seeding (Johnson, 1994). Spring wheat cultivars vary in their ability to suppress late germination and growth of foxtail (*Setaria spp.*) Such cultural practices consider life cycles of weeds and the crop, specific growth habits, variations in plant morphology and physiology, and environmental influences that affect weed species differently than the crop. Cultural manipulations must also consider soils and climate of the area; crops and crop rotation to follow; soil fertility and seedbed preparation practices; seeding date, rate, row spacing, and methods; competitive and allelopathic effects of weeds on the crop cultivar; production practices superimposed on the crop such as mowing, cultivating, harvesting, and grazing.

Biological weed control uses insects, plant diseases, nematodes, microorganisms, plants, animals, fish, or birds to reduce population or vigor of a weed species. Biological methods are the only economical or practical weed control in low-value rangelands (Julien, 1987). When suitable and effective agents are available, biological control is usually cheap and permanent with no treatment repetition. Yet there is no obvious biological control of weeds in arable cropland. Because weeds can be a vector for pests in arable crops, biological control can be a liability for pest release and damage.

Herbicides can be classified as selective or nonselective to plants; soil or foliar applied; preplant (incorporated or not), preemergence, or postemergence (broadcast, banded over the crop row, or directed); and contact or systemic. They also may be classified by chemical family or mode of action. Chemical weed control has rapidly become accepted as the major weed control method in developed countries, but environmental concerns are prompting reevaluations of herbicide use. Yet herbicides provide an opportunity to use tillage for other purposes such as the organic-carbon accretion mentioned earlier.

Integrated weed control is the most feasible approach for conservation production systems since no control method should be considered or relied on by itself. Each control method will provide some measure of weed inhibition, and the sum total of these various methods would provide dependable and adequate control of the composite of weeds present in a given conservation tillage field.

Integrated Weed Control Systems in Conservation Tillage

Conservation tillage systems as practiced today would not have been possible before the development of herbicides and their application technology. In conservation tillage systems, herbicides are integrated with tillage; in conventional crop production systems, herbicides supplement tillage. However, herbicide use is similar in all tillage systems that use full-width tillage (ERS, 1994) including those that have 30% surface cover after planting. As tillage is reduced, the entire pest complex is changed, and this may increase or decrease problems with weeds, insects, plant diseases, and nematodes. Weed species diversity may decrease, but the numbers of individual species may increase. Thus, farmers must identify and evaluate weed escapes and change their herbicides and other weed management methods accordingly.

Herbicide consistency and effectiveness is more critical in conservation tillage. Utility in conservation tillage requires incorporation without excessive tillage (e.g., double tandem disking) and formulations not subject to retention on plant residue above the soil surface. Early preplant herbicide treatments or *burn-down* herbicides are preferred because they may eliminate the need for preplant incorporation of herbicides. Weed control is best when herbicides are applied before weed emergence or when weeds are less than 3 cm tall.

When weed or weed-seed densities are low, postemergence herbicides with crop selectivity may replace soil-applied herbicides. Cultural and mechanical weed control methods can then be utilized initially and followed by postemergence application should weeds escape the nonchemical control. Weed seed banks may be monitored (Lybecker et al., 1991) for strategy decisions. When contact or translocated herbicides are nonselective (for control of various weed species), tillage for seedbed preparation can be more flexible. The ideal herbicide should destroy a broad spectrum of weed species initially and then exhibit enough soil residual to control repeated flushes of weed species during at least the first month following planting. Combinations of herbicides and both soil- and foliar-applied herbicides may be utilized to achieve acceptable weed control in conservation tillage systems, in which case persistence in the soil should be long enough for weed control until crop canopy closure. Thus, residual, translocated, and contact herbicides might be used in various cropping systems.

This detail of integrated weed control strategies indicates that weed control is already an advanced technology for crop production in adopted conservation tillage systems. There is still a serious need for controls on herbicide losses into the environment.

Herbicide Resistant Plants

Herbicide resistant weeds (HRW) are an emerging major problem to crop production, but herbicide resistant crops (HRC) represent a significant advance in weed management technology. HRW have become widely recognized by scientists and growers since the initial report (Ryan, 1970) of triazine resistance in common groundsel (*Senecio vulgaris* L.). Over 120 weed species have shown resistance in over 15 herbicide families such as aryloxyphenoxypropionic, bipyridilium, dinitronanilines, phenoxy, substituted urea, sulfonylurea, and triazine herbicides, and the number of recognized HRW is increasing annually (LeBaron and Gressel, 1982). Numerous scientific meetings and symposia concerning HRW have increased awareness about herbicide resistance, identified additional examples of resistance, researched the mechanisms of herbicide resistance, and exchanged ideas about practical solutions to HRW.

Numerous management strategies for avoiding or managing HRW have been proposed and are being used (Gunsolus, 1993). These strategies include herbicides use only when necessary, rotating herbicides to achieve different mechanisms of action, applying herbicide mixtures with different physiological sites of action, rotating crops as well as herbicides, utilizing both herbicide and alternative weed control methods, scouting fields for weed escapes, altering weed technology to control weed increasers, cleaning tillage and harvesting equipment between fields, evaluating the factors that facilitate development of HRW, and utilizing HRC as appropriate. Manufacturers of herbicides also are changing label recommendations/publications, and educating about use of herbicides that have a higher risk of selecting for HRW.

An alarming development of HRW is a multiple-resistance to chemically unrelated herbicides (Heap and Knight, 1986). Multiple-resistance to herbicides threatens crop production in situations where alternative methods of weed management are not generally available. Thus, the common recommendation of rotating herbicides with different modes of action may not solve the problem of multiple-resistance. This situation makes the development of alternative weed management technology even more important.

HRC have been used extensively during the past five decades, because crop tolerance is necessary for an herbicide to be selective (Burnside, 1992). In the past, a large number of organic compounds were screened against a limited number of major crop and weed species to discover HRC. Now a single herbicide can be evaluated in tissue culture for crop selectively against a large number of biotypes of a single crop, or sometimes gene transfer can be used to produce tolerant genotypes. Thus, one can now choose an environmentally *benign* herbicide and search for or develop resistant crop cultivars. Technology is now available and development is underway to produce HRC for many of our crops—even minor crops (Duke, 1995).

Both benefits and risks are involved in HRC. The advantages of these new HRC are: (a) more options for difficult weed management problems, (b) grower propensity to use integrated weed management when alternative weed control procedures can back up failures of cultural or mechanical methods, (c) increased weed management options for growers of minor crops, (d) economic advantages to growers, and (e) increased use of more environmentally *benign* herbicides. Concerns about HRC include: (a) mandatory use of herbicides, (b) use of biotechnology to encourage herbicide use, (c) a HRC becoming a weed, (d) reduced ecological diversity, (e) more herbicide carryover by facilitating the use of environmentally *harsh* herbicides, (f) less competition to weeds, and (g) decreased use of alternative weed control practices. Some of these expressed concerns are adverse publicity and others are genuine risk. HRC can contribute to productivity and profitability of agriculture if used wisely in an integrated weed management program (Burnside, 1992). Development of HRC should be limited to those programs that utilize environmentally benign herbicides and to crop production situations where HRC can reduce crop injury, improve weed management, increase crop yields or quality, and have a favorable or neutral effect on the environment. The risks and benefits of each HRC must be evaluated and approved on a case-by-case basis.

Technology of Plant Disease Control

Plant disease control was an active technology already in the 1930 era. An earlier reference (Leighty, 1938) discussed crop rotation, along with burning of infested residue when necessary, for disease control. Green manuring in these rotations for disease control had mixed results (Cook and Baker, 1983), but black fallow generally provided disease control whenever used in some of these earlier rotations. Cultivar resistance to disease had already been used to control *Fusarium* wilts in flax and cotton (Bruehl, 1987). Coons (1937) reported the share of resistant cultivars to range from 10 to nearly 100% of cropland depending on the crop; the estimated need for resistance was about

10% of the planted cropland. Most cultivars now have multigenic or horizontal resistance (or tolerance) to a specific host-pathogen combination and only a few have monogenic or vertical resistance (Shaner, 1981). Horizontal resistance may allow some disease to occur because the host may have several components of resistance and there is natural variation in the pathogen, i.e., races of a pathogen. Vertical resistance usually does not last long because the resistance may be specific to only one race of the pathogen, and natural selection to produce new races of the pathogen is a never-ending phenomenon.

Plant disease control has always been a crucial technology for soil and water conservation because of the need for economic return without extreme risk, quality as well as quantity of harvested product, and a consistent supply of crop residue for maintaining soil quality (Cook and Baker, 1983). Moreover, there is an ongoing effort to estimate the hidden (no specific symptoms in the shoot) losses produced by the ever-present crown and root diseases (Cook and Baker, 1983). Yield losses due to common root rot (*Bipolaris sorokiniana*) in wheat are estimated to be as large as 10% (Bruehl, 1987). Another persistent and ubiquitous pathogen is *Pythium ultimum*, which parasitizes juvenile roots of wheat and many other crops (Cook et al., 1990). Crown and root diseases are highly interactive with plant stresses (e.g., water or nutrients) because always at least a fraction of the root is rendered inactive. Plant stress itself weakens the root against infection and disease progress. Plant stress frequently induces a root exudation that provides a food source for microbial activity; exudates can be a stimulus for the pathogen to produce an infection body.

Technology for plant disease control has become even more critical because of changes in crop rotation since 1950 and placement of crop residues on or near the soil surface in reduced and conservation tillage. The long crop rotations of the 1930 era with sod and forage crops have been replaced with short (2 or 3 crops) rotations and monocrop (Tables 7 and 8). Outside of the mandatory fallow-wheat producing regions, the practice of summerfallow with frequent cultivation has nearly disappeared as a means for disease and cultural control. These changes have drastically reduced the options for biological and cultural control. Kommedahl (1981) listed nearly 100 diseases on 36 crops in which crop rotation had been documented to give control at some location at some time. In a majority of these instances the host-crop interval was 2 or more years. Three common diseases—*Sclerotinia sclerotiorum* (white mold), *Rhizoctonia solani* (Rhizoctonia root rot), and *Thielaviopsis basicola* (black root rot)—require from 5 to 10 years of an immune or unrelated crop to prevent soil carryover. A 3-year ecofallow rotation of corn or sorghum seeded into winter wheat stubble followed by fallow and winter wheat is one of a few that has provided excellent disease control (Watkins and Boosalis, 1994).

Nearly all discussions about biological control of plant pathogens focus on crop residue because crop residues provide food and a place for most plant pathogens to live and reproduce (Cook et al., 1978; Boosalis et al., 1981; Kommedahl, 1981; Watkins and Boosalis, 1994). The most significant change associated with conservation tillage adoption (Tables 4 and 5) is the placement of crop residues on or near the soil surface. When crop residue is not incorporated deeper than 10 cm during primary tillage, secondary tillage may bury some of the residue and expose previously buried residue (Allmaras et al., 1996a). When tillage rotation is used, there are many different systems to maintain crop residue in these positions. Kommedahl (1981) surveyed 20 instances in which crop residue management had a major effect on disease control; 18 of these had recommended *plow down* or *deep burial* of infested residue. Such conflicting information requires a structured understanding of how reduced tillage may have changed the needs for biological control.

Cook et al. (1978) and Boosalis et al. (1981) discuss plant residues as a survival habitat for pathogens, factors affecting survival and growth of the pathogen, and the effects of the reduced tillage (maintaining plant residue on or near the soil surface) on the physical and chemical environment. All three are drastically changed by the residue management associated with reduced tillage. Crop residues are also an organic amendment which can have both stimulatory and toxic effects on the surviving pathogen (Cook et al., 1978; Boosalis et al., 1981).

Bruehl (1987) suggests examining passive or active possession as a means by which the pathogen colonizes crop residue to manage food supply. Pathogens with passive possession store all of their hyphae within a resistive resting structure, such as sclerotia or thick-walled oospores which can be widely dispersed awaiting a chance to parasitize the new host. Other pathogens have active possession of the host and must maintain at least some metabolism as they remain in the dead plant material remaining on the surface or incorporated into the soil. Some of these pathogens cannot survive when they lose their pioneer status as residue decomposes, others can survive as a saprophyte through active metabolism within any colonized organic debris in the soil. *Cephalosporuim graminareum* is a good example of the former while *Pythium ultimum* typifies the later; the former is a wheat disease controlled by summerfallow while the latter readily survives summerfallow.

Watkins and Boosalis (1994) chose to separate the foliar (fungal and bacterial) diseases from the root and crown diseases. Tillage or lack of tillage has a more straightforward effect on survival, growth, and multiplication of foliar pathogens, especially the fungal pathogens that infect leaf, stem, head, and grain parts. These pathogens are strongly dependent on residue maintained on the surface or standing overwinter, awaiting dispersal of the inoculum during cool, wet springs. Often inoculum for foliar diseases can be dispersed into a field from outside sources. Burial of infested residue, even though only just below the surface to provide contact with soil microflora can give biological control—short crop rotations can also give control. Unfortunately fungicides are used frequently to control these diseases where cultivar resistance is poor, such as in *Septoria* diseases of small grains. Root and crown diseases require a much more diverse set of practices for biological control: crop sequence, fertilization, cultivar selection, separation of infested residue from the new root system (as in ridge tillage), residue decomposition rate, and amount and type of tillage operation. Inoculum for these diseases can remain in the soil for long periods through saprophytic colonization.

To continue the soil and water benefits from reduced and conservation tillage systems a multitude of approaches including rotational tillage and biological control will be needed to reduce the disease pressure on resistant or tolerant cultivars. Rotational tillage must involve timing as well as number of operations without moldboard plowing as more is learned about ecology of the pathogen while feeding and living within the crop residues. Since the positioning of crop residue is easily traced through knowledge of the tillage tool/system used (Staricka et al., 1991), it would be informative to clarify the pathogen ecology and control relationship afforded by the *plow down* discussed by Kommedahl (1981). Can this knowledge be used when all residue is retained within the upper 15 cm? The general increase in plant disease problems associated with adoption of reduced tillage indicates a need to strengthen and accelerate programs to develop resistant cultivars.

Shaner (1981) comprehensively listed resistance of major crops to disease, from which he concluded that disease resistance is a major control used in many crops especially for foliar diseases of annual crops. The number of diseases for which there is at least one commercially acceptable cultivar is large. Resistance for root diseases is far less than for crown and foliar diseases; part of the problem here is the greater ease to detect and recognize superior cultivars in foliar diseases. Because the development of disease resistant cultivars is a cumulative technology, progress on new cultivars is highly probable. Yet this germplasm improvement is expensive because other traits of production and quality must be maintained. Over the last 50 years this technology has provided many benefits because this mechanism for disease control is low cost and is automatically included by seed selection for the producing system. It is no small matter that this control method replace other disease control methods which could be adverse to soil and water conservation.

Crop Nutrition and Fertilizer Technology

Crop nutrition is a major technology for crop production, because it includes inputs or segments of : (a) nutrient uptake and accumulation in crops, (b) soil tests for adequacy of nutrients available to the crop, (c) tissue testing and visual observation of adequate but not excessive nutrient status of plants, (d) fertilizer formulation and blending, and (e) fertilizer application technology. All of these segments are included in the adopted technology because region, crop, and soil specificity associated with crop nutrition has required empirical studies wherever there was crop production. All of these segments of the crop nutrition technology will be needed to refine nutrient applications consistent with current plant needs. Crop nutrition and associated fertilization were among the first husbandry of crop production; the order of technology was a rudimentary understanding of plant nutrition in the seventeenth century followed by use of organic fertilizers and amendments in the eighteenth century, the beginning of inorganic fertilizer use in the nineteenth century, and finally the field experiments with chemical fertilizers (Nelson, 1990). Nelson (1990) briefly traced antecedents of fertilizer technologies through the Greek, Roman, Arabic, and Middle Ages writings or practices, but highlighted chemistry and plant growth sciences as the major advancements.

For crop production in the United States, Nelson (1990) discusses fertilizer use in three distinct 20-yr periods starting in 1920. Until 1940 the combined use of the three primary nutrients (N, P_2O_5, and K_2O) was indeed stagnated at 1.8 million or less tons, but consumption of all three nutrients increased roughly three- to five-fold in the 1940 to 1960 period. There was a continued modest growth in P_2O_5 and K_2O use in the 1960 to 1980 period while the use of N increased almost six-fold (Nelson 1990; ERS, 1994). Use of N, P_2O_5, and K_2O leveled off in 1980, and decreased somewhat since 1980 so that current use is about 12, 4, and 5 million tons, respectively. Fertilizer use in United States is now at least ten times greater than in 1940, when the major rate of use began. In the realm of fertilizer history, the 1980 to 1990 era reflected business trends and environmental concerns more than major changes in technology and use (Nelson, 1990).

To appreciate the need and impact of this annual fertilizer use, there are numerous estimates of the share of crop production provided by fertilization (Nelson, 1990). These estimates range from 30 to 55% depending on the period within the 1940 to 1980 era, but the most reliable estimate as of 1976 was 37%. Yield improvements from genetic improvement ranged from 30 to 80% depending on the crop (Fehr, 1984). The amount of fertilizer used over the 1980 to 1990 era does not reflect any impact of the major shift to conservation or reduced tillage since 1980, but the timing and application methods have changed.

Nearly all fertilizer in the 1920 to 1940 era was broadcast and incorporated by primary tillage, and there was a limited amount of band application (Salter, 1938). Fertilizer materials were dry powdered or pulverized, but granulated materials were in developmental stage. Salter (1938) lamented that the application equipment was not easily adjusted for rate of application and gave erratic application patterns; there was some application timing technology available but not being used. Schreiner and Anderson (1938) discussed the considerable soil tests (the available nutrient concept was already used) and the many plot tests to be used for generalized fertilizer recommendations. Now there is an array of application methods and timing, fertilizer carriers specific to nutrient element and form (gas, liquid, granules), and equipment for application (Randall et al., 1985; ERS, 1994).

Nutrient management practices, briefly summarized in Table 9, reveal the current status of fertilization and some of the expected trends in use (see ERS, 1994, for detail of practices since 1990). All of the five major segments of the crop nutrition technology are involved. Except for soybean, at least 85% of

Table 9. Nutrient Management Practices for Corn, Cotton, Soybean, Winter Wheat and Spring Wheat in 1993[a]

Fertilization Practice[b]	Corn	Cotton	Soybean	Winter Wheat	Spring Wheat
	% of planted crop				
Nutrient form applied					
Farm manure	18	4	6	3	4
Chemical fertilizer	97	85	29	86	87
Both	16	3	1	2	3
Soil tested	38	28	28	22	34
	% of fertilized crop				
Time of fertilizer appln.					
Before planting, fall	27	34	35	75	34
Before planting, spring	8	49	52	—	38
At planting	44	8	13	25	73
After planting	33	54	3	42	3
Application method					
Broadcast	71	55	90	54	33
Banded	42	24	9	21	74
Injected (knife action)	47	45	1	47	46

[a]From ERS, 1994.
[b]These practices summarized for major producing states—see Tables 4 and 5 for detail of states; except for soil tested, the sum of % may exceed 100 because more than one practice may be applied to a land parcel.

the planted crop received chemical fertilizer, and only 3% of the planted corn did not receive chemical fertilizer. Farm manure management, which complicates nutrient budgets of N (Meisinger and Randall, 1991) and P (Sharpley et al., 1994), is much more involved in corn than other crops. Even though much expert technology has been developed in soil testing and associated plant uptake and accumulation of nutrients in crops (see e.g., Barber, 1984; Marschner, 1986; Black, 1993), only about 30% of the planted crop was associated with soil testing (Table 9). Further elaboration on this issue (ERS, 1994) indicates that at least 70% of the soil tests involved nitrogen and that roughly 80% applied the nitrogen rate recommended by the soil test.

The time of fertilization and the method of application are combined to make fertilizer available to the plant when needed. Environmental safeguards must also be maintained. For corn, cotton, and soybean that is fertilized, about 35% of the preplant fertilization is in the fall; but the sum of timing greater than 100% indicates that preplant fertilization is a component of multiple applications (Table 9). For spring wheat, the fall preplant is about 50% of the preplant fertilization, which likely reflects less fear of NO_3^- leaching when anhydrous ammonia is fall applied in the colder and drier climate. The practice of banded fertilizer at planting of corn and wheat uses a well researched technology to strategically place the primary nutrients so they are accessible during early growth. The adopted conservation tillage system, with crop residue in the 0 to 10-cm zone (Tables 4 and 5), requires even more precise placement next to the seed to guard against nutrient availability to microbes active in residue decomposition—the main effect here is to delay the mineralization until later when the root system can access the nutrient (Randall et al., 1985). The small use of banding at or after soybean planting is a reflection that fertilization of the alternating corn or wheat provides ample residual for the soybean. The high use of injection (knife action) is curious (Table 9) because injection is designed for liquid and gaseous forms of N, high speed application, and for preplant-broadcast and postplant applications not close to the plant. For spring wheat the method could be used preplant to place the N below the crop residue concentrated in the 0 to 10-cm layer, because only 3% of the fertilization is postplant (Table 9).

This brief discussion demonstrates that the crop nutrition and fertilizer technology has advanced remarkably since 1940 and has transformed crop production. Crop nutrition has become specific for crop species, soil, past fertilization (and residual nutrient), crop rotation, climate or weather history, and tillage system. However, the multitude of factors influencing nutrient uptake and plant development (response), the empirical nature of the science involved, and the changed crop residue positioning in conservation tillage—all indicate a continued focus on technology for reliable control of crop response. Added to crop response is the need to fertilize consistent with retention of the agrichemical and plant nutrient within the target environment.

Nitrogen is a difficult nutrient to manage because of ammonia volatilization, denitrification, leaching losses, and the dynamic of the soil storage pool (Meisinger and Randall, 1991). A typical nitrogen cycle characterizes a difficult dynamic of the soil storage pool (National Research Council, 1993). All four management factors can be influenced by conservation tillage and the manner of fertilizer application. Control of leaching losses is a major constraint on N management; the control difficulty is reflected in the large amount of research and analysis typified by Follett et al. (eds, 1991) and National Research Council (1993). Potassium presents some difficulty for control in crop nutrition because rooting restrictions may cause a deficiency not identified in soil test (Randall et al., 1985), and most of these occur in reduced tillage systems.

Phosphorus presents a different problem because it is not mobile in the soil itself, only small amounts of the soil reservoir of P are available to plants, and it readily moves away from the target site into aquatic environments (National Research Council, 1993; Sharpley et al., 1994). Phosphorus may be dissolved in runoff or bound to the eroded sediment and organic matter. Judging from the high soil tests observed in 1989 (Sharpley et al., 1994) and bioavailability (to the eutrophication process) increase as soil test P increases, the application rate must be decreased and the method of application directed to maximum plant use, such as with banding at planting. The major shift in tillage away from moldboard plowing (Tables 4 and 5) and the emphasis on P application within the upper 15 cm (Table 9) can both contribute to the higher soil test P as discussed by Sharpley et al. (1994) for no-till systems. The residual P and the crop residue are being mechanically mixed in the upper 15 cm instead of the upper 30 cm, as was done when the primary tillage was done with a moldboard plow. Because the band supplement of P is directed at plant availability within the upper 15 cm, a soil test would be more rational if it were confined to the soil zone impacted by primary tillage, but this change of testing would require a more complete tillage history at least for the last 3 years.

A mass balance was suggested to compare fertilization inputs and crop harvest removal of nutrients (National Research Council, 1993). Application of this method since 1990 for corn, wheat, and cotton (ERS, 1994) shows average N balances uniformly positive for corn and cotton, but sometimes negative for wheat. Nitrogen inputs exceeded harvest output by 25% or more on more than 40% of the planted cotton and corn. Nitrogen inputs relative to outputs in wheat were much lower than in corn (both absolute balance and a 25% or more excess) except for TX and OK, where the input excess was similar to corn. Phosphorus inputs relative to harvested output were much less than for N; the 25% or more excessive P inputs were much higher in corn than wheat. The excessive P input in corn- and wheat-producing states (ERS, 1994) shows a distribution similar to the much higher soil test P in corn-producing vs. wheat-producing states (Sharpley et al., 1994).

Environmental Quality of Surface and Groundwater

The adoption of chemical technologies for pest control and crop nutrient supply has increased the potential for escape of agrichemicals into surface and groundwaters, not only because of the overall mass used but also because of the large proportion of the planted crop that receives agrichemical applications. Although the overall mass of herbicide use has declined somewhat since 1984, at least one herbicide was used on about 97% of the soybean, corn, and cotton planted in 1993 (ERS, 1994). About 50% of the planted wheat receives a herbicide—overall use in spring wheats ranges near 90% of the planted wheat whereas use in winter wheat is 30%. Insecticide use on potato, cotton, and corn constitutes 92, 68, and 30% respectively, of the planted crop (ERS, 1994). Overall mass of applied insecticide has also declined since 1980. The total use of P and K fertilizers has decreased somewhat since a peak use in 1980, but the use of N fertilizer has increased somewhat since 1987 after an erratic period of use increases and declines in the period 1980 to 1987. Use of at least one fertilizer on corn occurs on about 97% of the planted corn—uses in cotton, potatoes, soybean, wheat are about 85, 100, 30 and 80%, respectively (ERS, 1994). The intensity of agrichemical use in the production of corn is reflected in that nearly 43 to 47% of all agrichemicals, collectively, are used for corn production.

Conservation tillage associated with improved crop residue management has reduced the potential for agrichemical movement off-site, especially where there has been good subsoil drainage (Fawcett et al., 1994); the method of agrichemical application was also a critical factor. Conservation tillage has not increased the use of fertilizers and only in special circumstances has there been an increased demand for pesticides. In fact, ridge-till has permitted a distinct reduction in herbicide use because of the combined use of cultivation and herbicides (ERS, 1994).

A brief discussion of the mechanisms/pathways for agrichemical losses as related to erosion and sediment loss into surface waters, as well as runoff and infiltration/leaching displays complexity of the problem. Technology development and application must become more specific to the tillage system, crop, soil, and climate/weather because agrichemicals are used on nearly all of cultivated cropland.

Erosion and Sediment Losses to Surface Water

Conservation tillage has been adopted primarily because of economics and yield increases, but there have been remarkable reductions of topsoil loss. Surface residue cover, incorporated residue within top 10 cm, and random roughness are the tillage factors for control of soil erosion by water (Laflen et al., 1985). There is a direct relationship between erosion and the coverage of surface soil by crop residue. In general, every 10% of additional surface residue cover results in a 20% reduction in erosion, and 30% surface residue cover results in half the erosion that occurs from conventionally tilled soil (Moldenhauer, 1985). Conservation tillage can produce reductions in soil loss ranging from 7 to 22 t/ha per year compared to moldboard plowed land (Mannering et al., 1987). These reductions are due to improved erosion control, as well as lower runoff volumes and lower concentrations of suspended sediment in runoff (Gold and Loudon, 1989). Thus, conservation tillage has been widely promoted as an effective practice for reducing upland erosion and sediment delivery to surface waters.

Runoff and Infiltration

Moldboard plowing is often associated with oxidation of soil organic matter, burial of crop residue, destruction of surface macropores, and ultimate reductions in soil aggregation. Consequently, rates of runoff from moldboard tilled fields are typically greater than rates from fields managed with conservation tillage. Ellis et al. (1985) reported runoff losses from chisel plowed, flat, tile-drained fields that were 60% of the losses from similar fields managed with moldboard plowing. On steep fields with coarse-textured soils, the differences in runoff were even greater, with no runoff from conservation tilled fields for eight years during the growing season (Ellis et al., 1985). Baker (1987) reviewed several studies of runoff losses under natural rainfall, and showed that conservation tillage produced runoff losses ranging from 1 to 22% of the losses under conventional tillage, with the lower percentages associated with no-tillage. Some form of tillage, as with chisel plow or ridge till, prevented the large variation in herbicide runoff shown with strict no-till (Fawcett et al., 1994). Based upon these results, conservation tillage generally results in long-term reductions of runoff volume compared to conventional tillage, and the extent of reduction increases with the amount of crop residue maintained at the soil surface.

Reductions in runoff for conservation tillage compared to conventional tillage are generally attributed to three factors. The first is that residue left at the surface by conservation tillage intercepts raindrops, thereby preventing formation of surface crusts that retard infiltration. Secondly, rill erosion is reduced by miniature dams produced by the surface and partially buried residue. The third is that soil surface macropores are less abundant and more poorly connected to the subsurface under conventional tillage than conservation tillage, causing lower saturated hydraulic conductivities and infiltration rates under conventional tillage. Annual moldboard tillage buries the recently harvested residue and returns the one-year-old crop residue to the surface after the major share of decomposition has occurred (Allmaras et al., 1996a). Indirect evidence for increased infiltration under conservation tillage versus conventional tillage has been provided by Gold and Loudon (1982, 1989), who measured from 6 to 30% more drainage from subsurface tiles in a watershed managed with chisel tillage versus one managed with moldboard tillage. Another factor controlling infiltration is the use of primary tillage tools (other than the moldboard plow) that always retain crop residues and their decomposition products within the upper 10 cm, where polysaccharides may induce water stable aggregation (Allmaras et al., 1996b).

Surface and Groundwater Quality

Within the last decade, there has been a national interest to understand and manage the effects of conservation tillage on losses of nutrients and pesticides to surface and groundwaters. Conservation tillage typically reduces erosion and runoff, while increasing infiltration and drainage compared to conventional tillage systems based upon the moldboard plow (Gold and Loudon, 1989). Reduced sediment delivery from conservation tillage systems typically is associated with significant reductions in delivery of nutrients and pesticides that are adsorbed to soil particles. Examples of such reduction have been shown to occur with particulate phosphorus (Baker, 1985; Sharpley et al., 1994;) total Kjeldahl nitrogen (Baker, 1985; Ellis et al., 1985), and sediment bound pesticides (Baker, 1985). The reductions in nutrient and herbicide losses with sediment are typically not as great as the reduction in sediment loss, due to preferential erosion of finer-textured soil particles with high sorption capacities (Bailey et al., 1985).

Conservation tillage is less effective at controlling losses in runoff of soluble nutrients and weakly to moderately sorbed pesticides than at controlling losses of nutrients and pesticides that are bound to soil particles. The main reason is that reductions in runoff volume with conservation tillage are not as great as the reductions in sediment loss when compared to conventional tillage. Another factor is that concentrations of nutrients and herbicides in surface soil (within 5 cm of the surface) are often greater under conservation tillage than in conventional tillage (Kells and Meggitt,1985; Dick and Daniel, 1987; Fawcett et al., 1994). When runoff occurs, the water interacts with this surface zone, causing desorption of nutrients and pesticides. Runoff has also been observed to remove soluble phosphorus from crop residues on the surface(Timmons et al., 1970). In many cases, higher concentrations of soluble phosphorus and pesticides have been observed in runoff from conservation versus conventional tillage systems (Baker, 1985; Kenimer et al., 1987; Sharpley et al., 1994), although runoff volumes are reduced under conservation tillage.

Losses in runoff of soluble nutrients and pesticides with low to moderate organic carbon partition coefficients are typically greatest in the first runoff event following agrichemical application (Leonard et al., 1979; Glenn and Angle, 1987; Fawcett et al., 1994). While conservation tillage systems typically reduce runoff volumes over a long time period relative to conventional tillage systems, they may not always result in reduced runoff from storm events that occur early in the growing season. In such cases, the total annual loss of soluble nutrients and pesticides in runoff under conservation tillage may not be significantly less than annual losses under conventional tillage.

Leaching Losses of Nitrate

Leaching losses of nitrates in drainage water depend upon the volume of water leached through the soil profile and the concentration of nitrate-nitrogen in the leachate. Infiltration and drainage are typically increased with conservation tillage (especially for no-till) relative to conventional tillage. The timing of precipitation or irrigation relative to tillage and agrichemical application, the density and connectivity of surface and subsurface soil macropores, and the method of agrichemical application are each factors that influence the concentration of nitrates in leachate. Concentration of nitrogen compounds in soil also depends upon factors such as rate of application and extent of incorporation, extent of transformation of the parent compound (e.g., denitrification of N fertilizer), and extent of loss pathways by volatilization, runoff, or erosion.

In view of the many factors that influence agrichemical leaching losses, the relative effects on leaching losses of conservation tillage versus conventional tillage are site- and time-specific—this applies to nitrates.

To illustrate the complexity in interpreting effects of conservation tillage on leaching losses, consider that nitrate-nitrogen leaching losses under conservation tillage have been observed to be equal to (Kitur et al., 1984; Gold and Loudon, 1989), greater than (Tyler and Thomas, 1979), or less than (Kanwar et al., 1985) losses under conventional tillage. Typically, when leaching losses under conservation tillage are greater than those under conventional tillage it is because surface-applied agrichemicals have not been incorporated or leached into the soil matrix prior to the onset of macropore flow. When a solute is mixed into the soil matrix by low intensity rainstorms, the subsequent occurrence of intense storms and macropore flow leads to large volumes of macropore flow without delivery of solutes from the matrix region into macropores (Shipitalo et al., 1990; Edwards et al., 1992). Therefore, the timing and intensity of rainfall following tillage and agrichemical application influence the relative magnitudes of leaching loss under conservation versus conventional tillage (Granovsky et al., 1993).

PRECISION FARMING TO OPTIMIZE ENVIRONMENTAL QUALITY AND CROP PRODUCTIVITY

Precision farming or site-specific management has great potential for protection of water quality by preventing overapplication of agrichemicals in areas susceptible to leaching or runoff losses. Several recent workshops (Robert et al., 1993, 1995) indicate that variable rate technology (for application of agrichemicals) is already an adopted technology. For example, Kitchen et al. (1995) discuss the use of yield maps, landscape form, soil maps, and stochastic climate information to reduce the overall use of N fertilizer, reduce site-specific leaching potential, and yet improve the whole field yield of corn. Recent research focus is on the use of variable rate technologies in farm fields to provide nonpoint environmental protection (Larson et al., 1997). The environmental benefits of precision herbicide management were documented by Khakural et al. (1994), who showed a 22% reduction in alachlor concentrations in overland flow for soil-specific herbicide applications as compared to uniform applications. Mulla and Annandale (1990) found that roughly 42% of an applied bromide tracer was lost by leaching in a 57-ha irrigated potato field, and that reductions in leaching losses could be obtained by dividing the field into leaching risk zones based upon irrigation depth and soil texture. Mallawatantri and Mulla (1996) found that the probability of nitrate-N leaching losses in an irrigated potato farm were very high in 0.4 ha, high in 1.8 ha, moderate in 8.7 ha, low in 23.0 ha, and none in 23.1 ha. Thus, small regions of the 57-ha field contributed most of the nitrogen leaching loss.

There is great potential for the development of precision farming techniques to control erosion through variable tillage and residue cover in response to spatial variations in slope steepness and internal drainage. Present methods for conservation tillage treat whole fields uniformly, without regard for variations in erosion potential or limitations caused by soil wetness or internal drainage. Variable tillage could be achieved by engaging chiseling and disking tillage tools across the landscape at variable depths and angles of tillage (Voorhees et al., 1993).

While tillage is primarily important for seedbed preparation, weed control, planting, or fertilizer and pesticide incorporation/placement (Voorhees et al., 1993), conservation tillage which maintains crop residues at or just beneath the soil surface is often effective at increasing infiltration of water and reducing erosion of sediment (Langdale et al., 1994). Residue cover reduces water erosion because it intercepts and dissipates the energy of raindrop impact and reduces shear forces associated with flowing runoff water. Tillage may also produce soil surface roughness that temporarily increases surface water detention and reduces erosion.

The effectiveness of conservation tillage at controlling erosion through enhancements in surface roughness and residue cover are widely accepted. These benefits to soil conservation are offset to some degree, however, by detrimental effects of conservation tillage in soils with poor drainage and high moisture contents (Griffith and Mannering, 1985; Allmaras et al., 1991; Griffith and Wollenhaupt, 1994). Poorly drained soils are often characterized by excess moisture in spring, which can delay planting and reduce crop yield. This is especially problematic in northern latitudes, where short growing seasons prevail (Griffith and Mannering, 1985).

In subhumid climates where soils may be excessively wet in spring or late fall seasons, growers may produce serious soil compaction when they traffic wet and poorly drained soils. Reduced internal drainage can be permanent enough to decrease soil productivity. When such soils are bare, they dry more rapidly than similar soils with heavy residue cover. Thus, shallow tillage with chisels or disks on poorly drained soils may cause unwanted further delays in planting or lead to additional compaction problems for growers unwilling to wait for proper soil drying before tillage and planting. Triplett and van Doren (1985) found that reduced tillage practices on poorly drained soils gave acceptable corn yields when grown in a rotation, ridge tillage systems were employed, and tile drains were installed to improve soil drainage.

Tillage systems for controlling erosion while maintaining crop productivity depend upon many factors, which are both spatially and temporally varying within a given field. These include climate, cropping system, type and timing of tillage, amount, type, and depth of incorporated residues, slope steepness, internal soil drainage and moisture status, and ability to maintain adequate soil fertility and pest control. In fields with spatially varying soil conditions of drainage and moisture status, site-specific variation of tillage tool, depth, and angle has been advocated (Voorhees et al., 1993) in order to optimize soil conservation and water quality goals, while maintaining crop productivity. Where poorly drained soils may limit crop production, deeper tillage practices that bury a significant amount of residue can be imposed. Erosion from portions of the field with better drainage and steeper slopes can be controlled by site-specific chiseling or no-tillage. Precision conservation tillage may become a viable conservation tool throughout much of the midwestern U.S. because it allows conservation tillage to be applied selectively to the most erosive portions of the field while avoiding those portions where poor drainage limits crop emergence and trafficability.

SUMMARY AND CONCLUSIONS

Adoption of technology in crop production systems had its beginnings long before 1940, but the grand period of adoption began in earnest after the depression of 1930, and the end of World War II. Before 1930, the mechanical component had a start with the invention of many machines (small and often not durable) and tractors (powered with steam and internal combustion engines). After 1940 and up to the present, three components of technology (mechanical, biological, and chemical) were adopted simultaneously. Fortunately, the scientific approach not only produced a supply of technology but also the principles and practice of conservation (soil and water) when some of the ravishes of soil erosion became obvious in forestry and arable agriculture in the 1900 to 1940 period. Ruttan (1982) concisely summarized the transformation since 1940 from a resource-based to a science-oriented industry with simultaneous contributions from: (a) research to produce the new technology, (b) industrial sector to develop and market the new technology, and (c) farmers with their capacity to acquire and integrate the new technology. Another indicator is agricultural total output of 1.9% increase per year from 1948 to 1991, while the total input decreased 0.05% per year. Other sectors of the economy marvel at this accomplishment. Crop yield trends until 1940 were stagnated but after 1940 there are three distinct periods of a modest increase until 1960, a sharp increase from 1960 to 1980, and then a somewhat smaller increase since 1980. Soil and water conservation since 1980 improved as much or more than in any previous 15-year period, but adopted technology has intensified nonpoint source pollution of surface and groundwaters.

Our appraisal of technology adoption used resource data whenever possible. We evaluated the impact of adopted practices on soil and water conservation using process and results oriented research. Our focus was crop production systems and only selected technologies were discussed to trace their development and project future emphasis to preserve or improve soil and water conservation as well as environmental quality.

1. *Conservation tillage and crop residue management* are the cornerstone of

adopted technology and improved soil and water conservation (improved soil/water/air quality). Increased harvested production has provided more and more crop residue proportional to tabulated yield increase and estimated change in harvest index (Table 2). There is now almost complete farm use of tillage systems without the moldboard plow (Tables 4 and 5). Crop residue retained on or near the surface has reduced erosion nationwide about 30%; and improved carbon accretion has shifted agriculture from a net CO_2 producer to an accumulator of soil organic matter. The organic pool of plant nutrients in the soil has also increased. Improved water conservation and use undoubtedly reflect yet unmeasured infiltration increases due to more bioactivity near the soil surface.

Conservation tillage itself has changed since 1980 depending on the development of machinery (or mechanical), disease control, nutrient application, and weed control technologies. One component of the change may be tillage rotation to correct problems or capitalize on some gains only noticed after several years of a particular operation in the system. For instance, at least 3 or 4 years of comparative primary tillage are required to measure organic matter accumulation/loss.

2. *Machinery technology* has made many contributions throughout the 60-year period but the contributions to conservation tillage are notable, such as: uniform crop residue return by harvesters, tillage and seeding equipment to facilitate crop residue retention on or near the surface, fertilizer and pesticide applications equipment, hydraulic and other equipment to deliver engine power so that simultaneous multiple functions can be accomplished with one pass, and the accessories to monitor/control machinery function and performance. Machinery systems to provide precise zones of cultivation, seed placement, and agrichemicals placement is a critical component of improved crop production systems in the last 20 years. Machinery technology now has several major challenges to ease ground pressure and compaction, and provide machinery with computer interfaces needed for precision or site-specific farming.

3. *Crop rotation* has been urged upon farmers but resource information shows a high use of monoculture and short rotations characteristically different among various regions of the U.S. Conservation tillage adoption appeared not to be influenced by the regional variations of crop rotation. This dramatic change in crop residue management accomplished without the moldboard plow appears to substitute for the soil conservation objective in the pre-1950 crop rotations; crop rotation research using different tillage systems confirms this hypothesis. However, pest control problems have become more acute, the management of which requires a better knowledge of pest ecology in the crop residue related to placement in the soil.

4. *Weed management and control technology* has had and will continue to play a major role in the development of conservation tillage systems. Weed control has matured dramatically since 1980, when the primary focus had been herbicide action. Now there is a comprehensive approach consisting of

herbicide action, ecology of the weed and crop species, cultivation, persistence and movement of herbicides, and seed bank monitoring to anticipate weed populations. As other technologies change the nature of conservation tillage in site-specific farming, so also will weed control change. Weed control has challenges of its own with herbicide resistant weeds, the anticipated improvement of weed control in a herbicide resistant crop, and the development of herbicide use systems that reduce nonpoint source pollution of surface and groundwater.

5. *Plant disease control technology* through emphasis on pathogen ecology and the development of resistant cultivars has provided a vital role in crop production. In the monocrop and short rotation systems of some regions of the U.S., disease control has been the most critical impediment to conservation tillage; this is because most pathogens must colonize host residue or any crop residue to survive between infections of the host, and cultivar resistance is not reliable as the only method of disease control. Summerfallow as a cultural control has been reduced significantly in many regions as a component of crop rotation. To facilitate the development of disease control in conservation systems, research must emphasize germplasm development and epidemiology/ecology of the pathogen appropriate to the new paradigm of crop residue placement and economically feasible crop sequencing. Cultural and bio-control methods must also be expected for control.

6. *Crop nutrition and fertilizer technology* is a gigantic component of crop production systems; it includes crop nutrition and growth response/yield, soil and tissue testing, fertilizer formulation, and applications technology. To facilitate conservation tillage and associated change in crop residue placement, this technology was required to make major changes, and will be required to make even more adjustments to facilitate fertilizer placement and timing for efficient plant use and avoidance of nonpoint source pollution. Efficient use of N itself is a serious concern because only about 60%, at most, of applied N can be accounted for in the plant and soil. Both N and P are causing serious nonpoint source dangers to surface and groundwaters.

7. *Precision or site-specific farming* was a farm or land management practiced as in spot weed treatment or avoidance of tillage in erosion sensitive areas. Now it is an emerging technology that provides soil management and use of agrichemicals only where and in the amount needed. Within a field, these site managements may apply to soil types, leaching or runoff sensitive areas, weed infestations, sites with soil test high enough to reduce fertilization, soil internal drainage and merely sites with higher/lower yields than the mean. Site-specific farming is not a mere collection of controls on agrichemicals, because it is an integrated application of the technologies discussed here.

A true evaluation of technology adoption and the impact on conservation is much more than accounting for technology production and adoption in crop production, and predicted conservation benefit. Governmental actions, social causes within and outside of agriculture, and economics are involved. Forage production is also a vital part of the agricultural scene; the interested reader should consult a recent analysis of forage technology and its future (Wedin and Jones, eds., 1995).

The actors (research sector, industrial sector, farmers) involved in the transformation of agriculture into a science-oriented industry (Ruttan, 1982) will have an even greater challenge in the future to sustain production, improve soil and water conservation, and stem environmental degradation from nonpoint sources.

Acknowledgments

This article is published as Minnesota Agric. Expt. Station Publication 22525.

References

Adee, E.A., E.S. Oplinger, and C.R. Graw. 1994. Tillage, rotation sequence, and cultivar influences on brown stem rot and soybean yield. J. Prod. Agric. 7:341–347.

Allan, R.E. 1983. Harvest indexes of backcross-derived wheat lines differing in culm height. Crop Sci. 23:1029–1032.

Allmaras, R.R., S.M. Copeland, P.J. Copeland, and M. Oussible. 1996a. Spatial relations between oat residue and ceramic spheres when incorporated sequentially by tillage. Soil Sci. Soc. Am. J. 60:1209–1216.

Allmaras, R.R., S.M. Copeland, J.F. Power, and D.L. Tanaka. 1994. Conservation tillage systems in the northernmost central United States. pp. 255–284. In: M.E. Carter (ed.) Conservation Tillage in Temperate Agroecosystems. Lewis Publishers, Boca Raton, FL.

Allmaras, R.R., J.M. Kraft, and D.E. Miller. 1988. Soil compaction and incorporated crop residue effects on root health. Ann. Rev. Plant Path. 26:219–243.

Allmaras, R.R., G.W. Langdale, P.W. Unger, and R.H. Dowdy. 1991. Adoption of conservation tillage and associated planting systems. pp. 53–83. In: R. Lal and F.J. Pierce (eds.) Soil Management for Sustainability. Soil and Water Conserv. Soc., Ankeny, IA.

Allmaras, R.R., J.L. Pikul Jr., C.L. Douglas Jr., and R.W. Rickman, 1996b. Temporal character of surface seal/crust: Influences of tillage and crop residues. pp. 171–188. In: L.R. Ahuja and A. Garrison (eds.) "Real World" Infiltration, Colorado Water Resources Research Inst. Information Series 86. Fort Collins, CO.

Allmaras, R.R., J.L. Pikul, Jr., J.M. Kraft, and D.E. Wilkins. 1988. A method for measuring incorporated crop residue and associated soil properties. Soil Sci. Soc. Am. J. 52:1128–1133.

Allmaras, R.R., P.W. Unger, Jr., and D.E. Wilkins. 1985. Conservation tillage systems and soil productivity. pp. 357–412. In: R.F. Follett and B.A. Stewart (eds.) Soil Erosion and Crop Productivity. American Society of Agronomy, Madison, WI.

Angers, D.A., N. Bissonnette, A. Legere, and N. Samson. 1993a. Microbial and biochemical changes induced by rotation and tillage in a soil under barley production. Can. J. Soil Sci. 73:39–50.

Angers, D.A. and M.R. Carter. 1995. Aggregation and organic matter storage in cool, humid agricultural soils. pp. 193–211. In: M.R. Carter and B.A. Stewart (eds.) Structure and Organic Matter Storage in Agricultural Soils. Lewis Publishers, Inc., Boca Raton, FL.

Angers, D.A., N. Samson, and A. Legere. 1993b. Early changes in water-stable aggregation induced by rotation and tillage in a soil under barley production. Can. J. Soil Sci. 73:51–59.

Austin, M.E. 1981. Land Resource Regions and Major Land Resource Areas of the United States. Revised edition. Agric. Handbook 296. U.S. Dept. Agric., Washington, D.C.

Austin, R.B., J. Bingham, R.D. Blackwell, L.T. Evans, M.A. Ford, C.L. Morgan, and M. Taylor. 1980. Genetic improvements in winter wheat yields since 1900, and associated physiological changes. J. Agric. Sci., Camb. 94:675–689.

Babowicz, R.J., G.H. Hyde, and J.B. Simpson. 1983. Fertilizer effects under simulated no-till conditions. Paper number 83-1025, American Society of Agricultural Engineers Microfiche Collection, St. Joseph, MI.

Bailey, G. W., L. A. Mulkey, and R. R. Swank. 1985. Environmental implications of conservation tillage: A systems approach. pp. 239–265. In: F. M. D'Itri (ed.) A Systems Approach to Conservation Tillage. Lewis Publ., Chelsea, MI.

Baker, C.J. and K.E. Saxton. 1988. The cross-slot conservation-tillage grain drill opener. Paper number 88-1568, American Society of Agricultural Engineers Microfiche Collection, St. Joseph, MI.

Baker, J. L. 1985. Conservation tillage: Water quality considerations. pp. 217–238. In: F. M. D'Itri (ed.) A Systems Approach to Conservation Tillage. Lewis Publ., Chelsea, MI.

Baker, J. L. 1987. Hydrologic effects of conservation tillage and their importance relative to water quality. pp. 113–124. In: T. J. Logan, J. M. Davidson, J. L. Baker, and M. R. Overcash (eds.) Effects of Conservation Tillage on Groundwater Quality: Nitrates and Pesticides. Lewis Publ., Chelsea, MI.

Baker, J.L., T.S. Colvin, S.J. Marley and M. Dawelbeit. 1989. A point-injector applicator to improve fertilizer management. Appl. Eng. Agric. 5:334–338.

Barber, S.A. 1984. Soil Nutrient Bioavailability: A Mechanistic Approach. John Wiley & Sons, New York, NY.

Beauchamp, E.G. and R.P. Voroney. 1994. Crop carbon contribution to the soil with different cropping and livestock systems. J. Soil Water Conserv. 49:205–209.

Bidwell, R.G.S. 1974. Plant Physiology. Macmillan Publishing Co., Inc., New York, NY.

Black, C.A. 1993. Soil Fertility Evaluation and Control. Lewis Publ., Inc., Boca Raton, FL.

Blevins, R.L., R. Lal, J.W. Doran, G.W. Langdale, and W.W. Frye. 1998. Conservation tillage for erosion control and soil quality. pp. 51–68. In: F.J. Pierce and W.W. Frye (eds.) Advances in Soil and Water Conservation, Ann Arbor Press, Chelsea, MI.

Boosalis, M.G., B. Doupnik, and G.N. Odvody. 1981. Conservation tillage in relation to plant diseases. pp. 445–474. In: D. Pimental (ed.) Handbook of Pest Management in Agriculture, Volume I:445-474. CRC Press, Boca Raton, FL.

Bruce, R.R., G.W. Langdale, L.T. West, and W.P. Miller. 1992. Soil surface modification by biomass inputs affecting rainfall infiltration. Soil Sci. Soc. Am. J. 56:1614-1620.

Bruehl, G.W. 1987. Soil-Borne Plant Pathogens. Macmillan Publ. Co., New York, NY.

Burnside, O.C. 1992. Rationale for developing herbicide-resistant crops. Weed Technol. 6:621–625.

Buyanovsky, G.A. and G.H. Wagner. 1986. Post-harvest residue input to cropland. Plant Soil 93:57–65.

Buzzell, R.I. and B.R. Buttery. 1977. Soybean harvest index in hill-plots. Crop Sci. 17:968–970.

Carter, M.R. (ed.). 1994. Conservation Tillage in Temperate Agroecosystems. Lewis Publ., Inc., Boca Raton, FL.

Choi, C.H. and D.C. Erbach. 1986. Cornstalk residue shearing by rolling coulters. Trans. ASAE 29:1530–1535.

Christensen, L.A. and R.S. Magleby. 1983. Conservation tillage use. J. Soil Water Conserv. 38:156–157.

Ciais, P., P.P. Tans, M. Trolier, J.W.C. White, and R.J. Francey. 1995. A large northern hemisphere terrestrial CO_2 sink indicated by the $^{13}C/^{12}C$ ratio of atmospheric CO_2. Science 269:1098–1102.

Clark, R.L. and G.W. Langdale. 1990. Effects of combine traffic on soil compaction. ASAE Paper 90-1541. American Society Agricultural Engineering, St. Joseph, MI.

Cochrane, W.H. 1993. The Development of American Agriculture: A Historical Analysis. Second Edition. University of Minnesota Press, Minneapolis, MN.

Cook, R.J., C. Chamswarng, and W.-H. Tang. 1990. Influence of wheat chaff and tillage on *Pythium* populations in soil and *Pythium* damage to wheat. Soil Biol. Biochem. 22:939–947.

Cook, R.J. and K.F. Baker. 1983. The Nature and Practice of Biological Control of Plant Pathogens. American Phytopathology Society, St. Paul, MN.

Cook, R.J., M.G. Boosalis, and B. Doupnik. 1978. Influences of crop residues on plant diseases. pp. 147–163. In: W.R. Oschwald (ed.) Crop Residue

Management Systems. ASA Sp. Publ. 31. American Society of Agronomy, Madison, WI.

Coons, G.H. 1937. Progress in plant pathology: Control of disease by resistant varieties. Phytopath. 27:622-631.

Crookston, R.K. 1996. The rotation effect in corn. pp. 201–215. Proc. 50th Annual Corn and Sorghum Res. Conf., Chicago, IL.

Crookston, R.K., J.E. Kurle, P.J. Ford, and W.E. Lueschen. 1991. Rotational cropping sequence affects yield of corn and soybean. Agron. J. 83:108–113.

CTIC. 1991. National Survey of Conservation Tillage Practices, Including Other Tillage Types. Conservation Technology Information Center, West Lafayette, IN.

Culotta, E. 1995. Will plants profit from high CO_2? Science. 268:654–656.

Dick, W. A. and T. C. Daniel. 1987. Soil chemical and biological properties as affected by conservation tillage: Environmental implications. pp. 125–148. In: T. J. Logan, J. M. Davidson, J. L. Baker, and M. Overcash (eds.), Effects of Conservation Tillage on Groundwater Quality: Nitrates and Pesticides. Lewis Publ., Chelsea, MI.

Donald, C.M. and J. Hamblin. 1976. The biological yield and harvest index of cereals as agronomic and plant breeding criteria. Adv. Agron. 28:361–405.

Douglas, C.L. Jr., P.E. Rasmussen, and R.R. Allmaras. 1989. Cutting height, yield level, and equipment modification effects on residue distribution by combines. Trans. ASAE 32:1258–1262.

Douglas, Jr., C.L., P.E. Rasmussen, and R.R. Allmaras. 1992. Nutrient distribution following wheat-residue dispersal by combines. Soil Sci. Soc. Am. J. 56:1171–1177.

Duke, S.O. (ed.) 1995. Herbicide-Resistant Crops: Agricultural, Economic, Environmental, Regulatory, and Technological Aspects. Lewis Publishers, Boca Raton, FL.

Edwards, W. M., M. J. Shipitalo, W. A. Dick, and L. B. Owens. 1992. Rainfall intensity affects transport of water and chemicals through macropores in no-till soil. Soil Sci. Soc. Am. J. 56:52–58.

Ellis, B. G., A. J. Gold, and T. L. Loudon. 1985. Soil and nutrient losses with conservation tillage. pp. 275–298. In: F. M. D'Itri (ed.) A Systems Approach to Conservation Tillage. Lewis Publ., Chelsea, MI.

Erbach, D.E., J.E. Morrison, and D.E. Wilkins. 1983. Equipment modification and innovation for conservation tillage. J. Soil Water Conserv. 38:182–185.

ERS. 1994. Agricultural Resources and Environmental Indicators. Agric. Handbook 705. Economic Research Service, U.S. Dept. Agric., Washington, D.C.

Fawcett, R.S., B.R. Christensen, and D.P. Tierney. 1994. The impact of conservation tillage on pesticide runoff in surface water: A review and analysis. J. Soil Water Conserv. 49:126–135.

Fehr, W.R. (ed.) 1984. Genetic Contributions to Yield Gains of Five Major

Crop Plants. CSSA Spec. Publ. 7. Crop Science Society of America, Madison, WI.

Flint, M.L. and P.A. Roberts. 1988. Using crop diversity to manage pest problems: Some California examples. Am. J. Alternative Agric. 3:163–167.

Follett, R.F., D.R. Keeney, and R.M. Cruse (eds). 1991. Managing Nitrogen for Groundwater Quality and Farm Profitability. Soil Science Society of America, Madison, WI.

Forcella, F. and M.J. Lindstrom. 1988. Weed seed populations in ridge and conventional tillage. Weed Sci. 36:500–503.

Glenn, S. and J. S. Angle. 1987. Atrazine and simazine in runoff from conventional and no-till corn watersheds. Agric. Ecosys. Environ. 18:273–280.

Goering, C.E. 1989. Engine and Tractor Power. American Society Agricultural Engineers, St. Joseph, MI.

Gold, A. J. and T. L. Loudon. 1982. Nutrient, sediment, and herbicide losses in tile drainage under conservation and conventional tillage. Paper No. 82–2549, ASAE. St. Joseph, MI.

Gold, A. J. and T. L. Loudon. 1989. Tillage effects on surface runoff water quality from artificially drained cropland. Trans. ASAE 32:1329–1334.

Granovsky, A. V., E. L. McCoy, W. A. Dick, M. J. Shipitalo, and W. M. Edwards. 1993. Water and chemical transport through long-term no-till and plowed soils. Soil Sci. Soc. Am. J. 57:1560–1567.

Griffith, D. R. and J. V. Mannering. 1985. Differences in crop yields as a function of tillage system, crop management, and soil characteristics. pp. 47–57. In: F. M. D'Itri (ed.) A Systems Approach to Conservation Tillage. Lewis Publ., Chelsea, MI.

Griffith, D. R. and N. C. Wollenhaupt. 1994. Crop residue management strategies for the Midwest. pp. 15–36. In: J. L. Hatfield and B. A. Stewart (eds.) Crops Residue Management. Lewis Publ., Boca Raton, FL.

Gunsolus, J.L. 1993. Herbicide Resistant Weeds. North Central Regional Ext. Publ. 468, U.S. Dep. Agric., Washington, D.C.

Hammel, J.E., R.I. Papendick, and G.S. Campbell. 1981. Fallow tillage effects on evaporation and seedzone water content in a dry summer climate. Soil Sci. Soc. Am. J. 45:1016–1022.

Hanway, J.J. and C.R. Weber. 1971. Dry matter accumulation in eight soybean (*Glycine max* (L.) Merrill) varieties. Agron. J. 63:227–230.

Havlin, J.L., D.E. Kissel, L.D. Maddux, M.M. Claasen, and J.H. Long. 1990. Crop rotation and tillage effects on soil organic carbon and nitrogen. Soil Sci. Soc. Am. J. 54:448–452.

Heady, E.O. 1984. The setting for agricultural production and resource use in the future. pp. 8–30. In: B.C. English, J.A. Maetzold, B.R. Holding, and E.O. Heady (eds.) Future Agricultural Technology and Resource Conservation. Iowa State Univ. Press, Ames, IA.

Heap, I. and R. Knight. 1986. The occurrence of herbicide cross-resistance in a population of annual ryegrass, *Lolium rigidum*, resistant to diclofop-methyl. Aust. J. Agric. Res. 37:149–156.

Hood, C.E., R.E. Williamson, and G.J. Wells. 1991. Tractors for 2000. Agric. Eng. 5:22–25.

Horner, G.M., A.G. McCall, and F.G. Bell. 1994. Investigations in Erosion Control and Reclamation of Eroded Land at the Palouse Conservation Expt. Station, Pullman, WA, 1931–1942. USDA Tech. Bull. 860. U.S. Dept. of Agric., Washington, D.C.

Howell, T.A. 1990. Grain, dry matter yield relationships for winter wheat and grain sorghum—Southern Great Plains. Agron. J. 82:914–918.

Hyde, G., D. Wilkins, K.E. Saxton, J.E. Hammel, G. Swanson, R. Hermanson, E. Dowding, J. Simpson, and C. Peterson. 1987. Reduced tillage and seeding equipment development. pp. 41–56. In: L.F. Elliott, R.J. Cook, M. Molnau, R.E. Witters and D.L. Young. (eds). STEEP-Conservation Concepts and Accomplishments. Washington State Univ., Pullman.

Jensen, H.E., P. Schjonning, S.A. Mikkelsen, and K.B. Madsen (eds.) 1994. Soil Tillage for Crop Production and Protection of the Environment. Proc. 13th Intern. Conf. ISTRO. Volume I and II. Royal Veterinary and Agricultural University, Copenhagen, Denmark.

Johnson, R.C. 1994. Influence of no till on soybean cultural practices. J. Prod. Agric. 7:43–49.

Julien, M.H. (ed.). 1987. Biological Control of Weeds: A World Catalogue of Agents and Their Target Weeds. CAB International, Wallingford, Oxon, OX10 DE, United Kingdom.

Kanwar, R. S, J. L. Baker, and J. M. Laflen. 1985. Nitrate movement through the soil profile in relation to tillage system and fertilizer application method. Trans. ASAE 28:1802–1807.

Karlen, D.L., G.E. Varvel, D.E. Bullock, and R.M. Cruse. 1994. Crop rotations for the 21st century. Adv. Agron 53:1–45.

Karlen, D.L., N.C. Wollenhaupt, D.C. Erbach, E.C. Berry, J.B. Swan, N.S. Nash, and J.L. Jordahl. 1994. Long-term tillage effects on soil quality. Soil Tillage Res. 32:313–27.

Keeling, C.D., Whorf, T.P., M. Wahlen, and J. van der Pilcht. 1995. Interannual extremes in the rate of rise of atmospheric carbon dioxide since 1980. Nature 375:660–670.

Kellogg, R.L., G.W. TeSelle, and J.J. Goebel. 1994. Highlights from the 1992 National Resources Inventory. J. Soil Water Conserv. 49:521–527.

Kells, J. J. and W. F. Meggitt. 1985. Conservation tillage and weed control. pp. 123–129. In: F. M. D'Itri (ed.) A Systems Approach to Conservation Tillage. Lewis Publ., Chelsea, MI.

Kenimer, A. L., S. Mostaghimi, R. W. Young, T. A. Dillaha, and V. O. Shanholtz. 1987. Effects of residue cover on pesticide losses from conventional and no-tillage systems. Trans. ASAE 30:953–959.

Khakural, B. R., P. C. Robert, and W. C. Koskinen. 1994. Runoff and leaching of alachlor under conventional and soil-specific management. Soil Use Manage. 10:158–164.

King, C.A. and L.R. Oliver. 1994. A model for predicting large crabgrass

(*Digitaria sanguinalis* L.) emergence as influenced by temperature and water potential. Weed Sci. 42:561–567.

Kitchen, N.R., D.F. Hughes, K.A. Sudduth, and S.J. Birrell. 1995. Comparison of variable rate to single rate nitrogen fertilizer application: Corn production and residual soil NO_3-N. pp. 425–442. In: P.C. Robert, R.H. Rust, and W.E. Larson (eds.) Site Specific Management for Agricultural Systems. American Society of Agronomy, Inc., Madison, WI.

Kitur, B. K., M. S. Smith, R. L. Blevins, and W. W. Frye. 1984. Fate of N-depleted ammonium nitrate applied to no-tillage and conventional tillage corn. Agron. J. 76:240–242.

Kommedahl, T. 1981. The environmental control of plant pathogens using eradication. pp. 297–315. In: D. Pimental (ed.) Handbook of Pest Management in Agriculture, Volume 1. CRC Press, Boca Raton, FL.

Laflen, J.M., G.R. Foster, and C.A. Onstad. 1985. Simulation of individual storm soil loss for modeling the impact of soil erosion on soil productivity. pp. 285–295. In: S.A. El-Swaify, W.C. Moldenhauer, and A. Lo (eds.) Soil Erosion and Conservation. Soil and Water Conservation Society, Ankeny, IA.

Lal, R., A.A. Mahboubi, and N.R. Fausey. 1994. Long-term tillage and rotation effects on properties of a central Ohio soil. Soil Sci. Soc. Am. J. 58:517–522.

Lamb, J.A., G.A. Peterson, and C.R. Fenster. 1985. Wheat fallow tillage systems effect on a newly cultivated grassland's nitrogen budget. Soil Sci. Soc. Am. J. 49:352–356.

Langdale, G. W., E. E. Alberts, R. R. Bruce, W. M. Edwards, and K. C. McGregor. 1994. Concepts of residue management: Infiltration, runoff, and erosion. pp. 109–124. In: J. L. Hatfield and B. A. Stewart (eds.) Crops Residue Management. Lewis Publ., Inc., Boca Raton, FL.

Larson, W.E., C.E. Clapp, W.H. Pierre, and Y.B. Morachan. 1972. Effects of increasing amounts of organic residue on continuous corn. II. Organic carbon, nitrogen, phosphorus, and sulfur. Agron. J. 64:204–208.

Larson, W.E., J.A. Lamb, B.R. Khakural, R.B. Ferguson, and G.W. Rehm. 1997. Potential of site-specific management for nonpoint environmental protection. pp. 337–367. In: F.J. Pierce and E.J. Sadler (eds.) The State of Site-Specific Management for Agriculture. American Society of Agronomy, Miscellaneous Publication. Madison, WI.

Larson, W.E., M.J. Mausbach, B.L. Schmidt, and P. Crosson. 1998. Policy and government programs in soil and water conservation. pp. 195–218. In: F.J. Pierce and W.W. Frye (eds.) Advances in Soil and Water Conservation. Ann Arbor Press, Chelsea, MI.

Larson, W.E. and F.J. Pierce. 1991. Conservation and enhancement of soil quality. pp. 175–203. In: Evolution for Sustainable Land Management in the Developing World, Vol. 2: Technical Papers. International Board for Soil Research and Management. Bangkok, Thailand.

LeBaron, H.M. and J. Gressel (eds.) 1982. Herbicide Resistance in Plants. John Wiley & Sons, Inc., New York, NY.

Leighty, C.L. 1938. Crop rotation. pp. 406–430. In: Soils and Men, Yearbook of Agriculture. U.S. Dept. of Agric., Washington, D.C.

Leonard, R. A., G. W. Langdale, and W. G. Fleming. 1979. Herbicide runoff from upland Piedmont watersheds—Data and implications for modeling pesticide transport. J. Environ. Qual. 8:223–229.

Lybecker, D.L., E.E. Schweizer, and P. Westra. 1991. Weed management decisions in corn based on bioeconomic modeling. Weed Sci. 39:124–129.

Macy, T.S. 1994. Yield map generation in a custom harvester environment. Paper number 94-1581. American Society of Agricultural Engineers Microfiche Collection. St. Joseph, MI.

Magleby, R., C. Sandretto, W. Crosswhite, and C.T. Osborn. 1995. Soil Erosion and Conservation in United States: An Overview. Agric. Information Bulletin 718. Economic Res. Service, USDA, Washington, D.C.

Mahler, R.L. 1985. The effect of soil moisture on the tolerance of wheat to different rates and sources of starter N fertilizer. pp. 23–31. In: D. Hayes (ed). Proceedings. 36th Annual Northwest Fertilizer Conference, Salt Lake City, UT, 16–17 July 1985. Northwest Plant Food Association, Portland, OR.

Mahler, R.L., L.K. Lutcher, and D.O. Everson. 1989. Evaluation of factors affecting emergence of winter wheat planted with seed-banded nitrogen fertilizers. Soil Sci. Soc. Am. J. 53:571–575.

Mallawatantri, A. P. and D. J. Mulla. 1996. Uncertainties in leaching risk assessments due to field averaged transfer function parameters. Soil Sci. Soc. Am. J. 60:722–726.

Mannering, J. V., D. L. Schertz, and B. A. Julian. 1987. Overview of conservation tillage. pp. 3–18. In: T. J. Logan, J. M. Davidson, J. L. Baker, and M. R. Overcash (eds.) Effects of Conservation Tillage on Groundwater Quality: Nitrates and Pesticides. Lewis Publ., Chelsea, MI.

Marschner, H. 1986. Mineral Nutrition of Higher Plants. Academic Press Inc., San Diego, CA.

McCalla, T.M. and T.J. Army. 1961. Stubble mulch farming. Adv. Agron. 13:125–196.

McClellan, R.C. 1987. A comparison of soil seed zone moisture in conventional tilled fallow fields and chemical fallow fields in Whitman County. pp. 607–615. In: L.F. Elliott, R.J. Cook, M. Molnau, R.E. Witters and D.L. Young (eds). STEEP-Conservation Concepts and Accomplishments. Washington State University, Pullman.

Meese, B.G., P.R. Carter, E.S. Oplinger, and J.W. Pendleton. 1991. Corn/soybean rotation effect as influenced by tillage, nitrogen and hybrid/cultivar. J. Prod. Agric. 4:74–80.

Meisinger, J.J. and G.W. Randall. 1991. Estimating nitrogen budgets for soil-crop systems. pp. 85–124. In: R.F. Follett, D.R. Keeney, and R.M. Cruse (eds.). Managing Nitrogen for Groundwater Quality and Farm Profitability. Soil Science Society of America, Madison, WI.

Miller, F.P., W.D. Rasmussen, and L.D., Meyer. 1985. Historical perspective of soil erosion in United States. pp. 23–48. In: R.F. Follet and B.A. Stewart (eds.) Soil Erosion and Crop Productivity. American Society of Agronomy, Madison, WI.

Moldenhauer, W. C. 1985. A comparison of conservation tillage systems for reducing soil erosion. pp. 111–122. In: F. M. D'Itri (ed.) A Systems Approach to Conservation Tillage. Lewis Publ., Chelsea, MI.

Morrison, J.E., Jr. and K.N. Potter. 1994. Fertilizer solution placement with a coulter-nozzle applicator. Appl. Eng. Agric. 10:7–11.

Morrison, J.E., Jr., R.R. Allen, D.E. Wilkins, G.M. Powell, R.D. Grisso, D.C. Erbach, L.P. Herndon, D.L. Murray, G.E. Formanek, D.L. Pfost, M.M. Herron, and D.J. Baumert. 1988. Conservation planter, drill and air-type seeder selection guideline. Appl. Eng. Agric. 4:300–309.

Mulla, D. J. and J. G. Annandale. 1990. Assessment of field-scale leaching patterns for management of nitrogen fertilizer application. pp. 55–63. In: K. Roth, H. Fluhler, W. Jury, and J. Parker (eds.), Field-Scale Water and Solute Flux in Soils. Birkhauser-Verlag, Basel, Switzerland.

National Research Council. 1993. Soil and Water Quality: An Agenda for Agriculture. National Academy Press, Washington, D.C.

Nelson, L.B. 1990. History of the U.S. Fertilizer Industry. TVA, Muscle Shoals, AL.

Papendick, R.I., L.F. Elliott, and K.E. Saxton. 1985. Paired rows push no-till grain yields up. The Service 6(7):6–7. Soil and Water Conservation, U. S. Dept. of Agric. Soil Conservation Service. Washington, D.C.

Payton, D.M., G.M. Hyde and J.B. Simpson. 1985. Equipment and methods for no-tillage wheat planting. Trans. ASAE 28:1419–1424, 1429.

Pearson, L.C. 1967. Crop rotations, Chapter 5. pp. 73–84. In: Principles of Agronomy. Reinhold, New York, NY.

Peterson, C.L. 1991. A comparison of expert systems and simulation techniques for control of a fertilizer applicator. pp. 373–378. In: Proc. 1991 Symposium, 16–17 December 1991, Chicago, IL. American Society of Agricultural Engineers, St. Joseph, MI.

Peterson, C.L., E.A. Dowding, K.N. Hawley, and R.W. Harder. 1983. The chisel-planter minimum tillage system. Trans. ASAE 26:378–383, 388.

Pierce, F.J., W.E. Larson, and R.H. Dowdy. 1986. Field estimates of C factors: How good are they and how do they affect calculations of erosion. In: Soil Conservation—Assessing the National Resources Inventory. Volume 2. National Academy Press, Washington D.C.

Pieters, A.J. and R. McKee. 1938. The use of cover and green-manure crops. pp. 431–444. In: Soils and Men. Yearbook of Agriculture. U.S. Dept. Agric., Washington, D.C.

Pikul, J.L., Jr. and R.R. Allmaras. 1986. Physical and chemical properties of a

Haploxeroll after fifty years of residue management. Soil Sci. Soc. Am. J. 50:214–219.

Power, J.F. and R.F. Follett. 1987. Monoculture Sci. Am. 256:78–86.

Prihar, S.S. and B.A. Stewart. 1990. Using upper-bound slope through origin to estimate genetic harvest index. Agron. J. 82:1160–1165.

Prihar, S.S. and B.A. Stewart. 1991. Sorghum harvest index in relation to plant size, environment, and cultivar. Agron. J. 83:603–608.

Randall, G.W., K.L. Wells, and J.J. Hanway. 1985. Modern techniques in fertilizer application. pp. 521–560. In: O.P. Englestad (ed.), Fertilizer Technology and Use. Soil Science Society of America, Madison, WI.

Raper, R.L., A.C. Bailey, E.C. Burt, T.R. Way, and P. Liberati. 1995. The effects of reduced inflation pressure on soil-tire interface stresses and soil strength. J. Terramechanics 32:43–51.

Rasmussen, P.E., R.R. Allmaras, C.R. Rohde, and N.C. Roager, Jr. 1980. Crop residue influences on soil carbon and nitrogen in a wheat-fallow system. Soil Sci. Soc. Am. J. 44:596–600.

Rasmussen, P.E. and W.J. Parton. 1994. Long-term effects of residue management in wheat-fallow: I. Inputs, yield, and soil organic matter. Soil Sci. Soc. Am. J. 58:523–530.

Reeves, D.W. 1994. Cover crops and rotations. pp. 125–172. In: J.L. Hatfield and B.A. Stewart (eds.). Crops Residue Management. Lewis Publ., Inc., Boca Raton, FL.

Reicosky, D.C., W.D. Kemper, G.W. Langdale, C.L. Douglas, Jr., and P.E. Rasmussen. 1995. Soil organic matter changes resulting from tillage and biomass production. J. Soil Water Conserv. 50:253–261.

Renard, K.G., G.R. Foster, G.A. Weesies, and J.P. Porter. 1991. RUSLE: Revised universal soil loss equation. J. Soil Water Conserv. 46:30–33.

Ritchie, S.W. and J.J. Hanway. 1982. How a Corn Plant Develops. Special Report 48. Iowa State University, Ames, IA.

Robert, P.C., R.H. Rust, and W.E. Larson (eds.). 1993. Proc. Soil Specific Crop Management: A Workshop on Research and Development Issues. 14–16 April 1992, Minneapolis, MN. ASA, CSSA, SSSA, Madison WI.

Robert, P.C., R.H. Rust, and W.E. Larson (eds.). 1995. Site Specific Management for Agricultural Systems. 27–30 March 1994. ASA, CSSA, SSSA, Madison WI.

Rogers, H.H., G.B. Runion, and S.V. Krupa. 1994. Plant responses to atmospheric CO_2 enrichment with emphasis on roots and the rhizosphere. Environ. Pollut. 83:155–189.

Rudolph, W. 1994. Strategies for prescription application using the chemical injection control system with computer commanded rate changes. Paper number 94-1585 American Society of Agricultural Engineers Microfiche Collection. St. Joseph, MI.

Ruttan, V.W. 1982. Agricultural Research Policy. Univ. Minnesota Press, Minneapolis, MN.

Ryan, G.F. 1970. Resistance of common groundsel to simazine and atrazine. Weed Sci. 18:614–616.

Salter, R.M. 1938. Methods of applying fertilizers. pp. 546–562. In: Soils and Men. Yearbook of Agriculture. US Dept. Agric., Washington, D.C.

Schriener, O. and M.S. Anderson. 1938. Determining the fertilizer requirements of soils. pp. 469–486. In Soils and Men. Yearbook of Agriculture. US Dept. Agric., Washington, D.C.

Shaner, G. 1981. Genetic resistance for control of plant disease. pp. 395–444. In: D. Pimental (ed.) Handbook of Pest Management in Agriculture, Volume 1. CRC Press, Boca Raton, FL.

Sharpley, A. N., S. C. Chapra, R. Wedepoh, J. T. Sims, T. C. Daniel, and K. R. Reddy. 1994. Managing agricultural phosphorus for protection of surface waters: Issues and options. J. Environ. Qual. 23:437–451.

Shertz, D.L. 1988. Conservation tillage: An analysis of acreage projections in the United States. J. Soil Water Conserv. 43:256–258.

Shipitalo, M. J., W. M. Edwards, W. A. Dick, and L. B. Owens. 1990. Initial storm effects on macropore transport of surface-applied chemicals in no-till soil. Soil Sci. Soc. Am. J. 54:1530–1536.

Singh, I.D. and N.C. Stoskopf. 1971. Harvest index in cereals. Agron. J. 63:224–226.

Skidmore, E.L. 1994. Wind erosion. pp. 265–293. In: R. Lal (ed.) Soil Erosion Research Methods. 2nd Ed. St. Lucie Press, Delray Beach, FL.

Skidmore, E.L., J.L. Hagen, D.V. Armbrust, A.A. Durar, D.W. Freyrear, K.N. Potter, L.E. Wagner, and T.M. Zobeck. 1994. Methods for investigating basic processes and conditions affecting wind erosion. pp. 295–330. In: R. Lal (ed.) Soil Erosion Research Methods. 2nd Ed. St. Lucie Press, Delray Beach, FL.

Staricka, J.A., R.R. Allmaras, and W.W. Nelson. 1991. Spatial variation of crop residue incorporated by tillage. Soil Sci. Am. J. 55:1668–1674.

Staricka, J.A., R.R. Allmaras, W.W. Nelson, and W.E. Larson. 1992. Soil aggregate longevity as determined by incorporation of ceramic spheres. Soil Sci. Soc. Am. J. 56:1591–1597.

Stout, T.T., D.L. Forster, L.L. Labao, and R.D. Munoz. 1989. Organization and Performance of Ohio Farm Operations in 1986. Res. Bull 1185. Columbus: Ohio Agric. Res. and Dev. Ctr. Ohio State University, Columbus.

Tessier, S., G.M. Hyde, R.I. Papendick, and K.E. Saxton. 1991. No-till seeders effects on seed zone properties and wheat emergence. Trans. ASAE 34:733–739.

Timmons, D. R., R. F. Holt, and J. J. Latterell. 1970. Leaching of crop residues as a source of nutrients in surface runoff water. Water Resour. Res. 6:1367–1375.

Triplett, G. B. and D. M. van Doren. 1985. An overview of the Ohio conservation tillage research. pp. 59–68. In: F. M. D'Itri (ed.) A Systems Approach to Conservation Tillage. Lewis Publ., Chelsea, MI.

Tweeten, L. 1995. The structure of agriculture: Implications for soil and water conservation. J. Soil Water Conserv. 50:347–352.

Tyler, D. D. and G. W. Thomas. 1979. Lysimeter measurement of nitrate and chloride losses from conventional and no-tillage corn. J. Environ. Qual. 6:63–66.

USDA. 1982. RCA, A National Program for Soil and Water Conservation: 1982. Final Program Report and Environmental Impact Statement. USDA, Washington, D.C.

USDA 1965. Agricultural Statistics 1965. USDA, Washington, DC.

USDA 1977. Agricultural Statistics 1977. USDA, Washington, DC.

USDA 1994. Agricultural Statistics 1994. USDA, Washington, DC.

Vogel, O.A., R.E. Allan, and C.J. Peterson. 1963. Plant and performance characteristics of semidwarf winter wheats producing most efficiently in eastern Washington. Agron. J. 55:397–399.

Voorhees, W.B. 1992. Wheel-induced soil physical limitations to root growth. Adv. Soil Sci. 19:73–95. Springer-Verlag, New York, N.Y.

Voorhees, W. B., R. R. Allmaras, and M. J. Lindstrom. 1993. Tillage considerations in managing soil variability. pp. 95–112. In: P. C. Robert, R. H. Rust, and W. E. Larson (eds.) Proc. Soil Specific Crop Management: A Workshop on Research and Development Issues. ASA, CSSA, SSSA, Madison, WI.

Watkins, J.E. and M.G. Boosalis. 1994. Plant disease incidence as influenced by conservation tillage systems. pp. 261–283. In: P.W. Unger (ed.) Managing Agricultural Residues. Lewis Publ., Inc., Boca Raton, FL.

Wedin, W.F. and J. P. Jones. 1995. Innovative Systems for Utilization of Forage, Grassland, and Rangeland Resources. Proceedings of a Workshop. 22–24 Sept. 1993. Airlie, VA. Minnesota Extension Service, St. Paul, MN.

Wicks, G.A., O.C. Burnside, and W.L. Felton. 1994. Weed control in conservation tillage systems. pp. 241–244. In: P.W. Unger (ed.) Managing Agricultural Residues. Lewis Publ., Inc., Boca Raton, FL.

Wilkins, D.E., E.A. Dowding, G.M. Hyde, C.L. Peterson, and G.J. Swanson. 1987. Conservation tillage equipment for seeding. pp. 571–577. In: L.F. Elliott, R.J. Cook, M. Molnau, R.E. Witters and D.L. Young. (eds.) STEEP-Conservation Concepts and Accomplishments. Washington State University, Pullman.

Wilkins, D.E., F. Bolton, and K. Saxton. 1992. Evaluating seeders for conservation tillage production of peas. Appl. Eng. Agric. 8:165–170.

Wilkins, D.E. 1988. Apparatus for Placement of Fertilizer Below Seed with Minimum Soil Disturbance. U.S. Patent Number 4,765-263. Issued: 23 August.

Wilkins, D.E. and H.A. Haasch. 1990. Performance of a deep furrow opener for placement of seed and fertilizer. pp. 58–62. In: 1990 Columbia Basin Agric. Research. Special Rep. 860. Oregon Agricultural Experiment Station, Corvallis.

Wilkins, D.E. and J.M. Kraft. 1988. Managing crop residue and tillage pans for pea production. pp. 927–930. In: 11th Intern. Conf. Proc. Vol. 2. ISTRO. 11-15 July 1988, Edinburgh, Scotland.

Wilkins, D.E., C.L. Douglas, Jr., and J.L. Pikul, Jr. 1996. Header loss for Shelbourne Reynolds stripper-header harvesting wheat. Appl. Eng. Agric. 12:159–162.

Womac, A.R. and F.D. Tompkins. 1990. Probe-type injector for fluid fertilizers. Appl. Eng. Agric. 6:149–154.

Womac, A.R. and F.D. Tompkins. 1991. Rotary meter for sequential injection of corrosive fluids. Appl. Eng. Agric. 7:325–328.

Wych, R.D. and D.C. Rasmusson. 1983. Genetic improvement in malting barley cultivars since 1920. Crop Sci. 23:1037–1040.

Wych, R.D. and D.D. Stuthman. 1983. Genetic improvement in Minnesota-adapted oat cultivars released since 1923. Crop Sci. 23:879–881.

The Human Dimension of Soil and Water Conservation: A Historical and Methodological Perspective

Pete Nowak and Peter F. Korsching

Introduction

Although we have had over half a century of federal, state, and private soil and water conservation programs in the United States, the efficacy of these programs on reducing soil loss, reducing further soil quality degradation, and maintaining clean water often is questioned (US GAO, 1977; USDA, 1989). This last half century can be characterized by significant investment in the monitoring and modeling of degradation processes, design and cost-sharing the implementation of remedial practices, and developing new institutional arrangements to further the use of these remedial practices (Bromley, 1991). Nonetheless, erosion, and more recently the recognition of deterioration in soil quality (National Academy of Science, 1993) and its relation to water quality, continues to be of significant concern.

Slower to develop has been a recognition that a comprehensive understanding of agricultural natural resource management must include the beliefs, motives, and actions of the resource manager; that is, this body of science must include some understanding of the individuals and organizations who manipulate the resource base to achieve their production and profitability objectives. Any scientific analysis of accelerated soil erosion processes must be more than some combination of soil science, agronomy, engineering, accounting, and plant biology—it must incorporate the land user or farmer as a central component of this analysis. Simply put, a scientific understanding of erosion processes and their application to the development of "technical fixes" does not constitute a solution to soil erosion problems. All the good intentions of science and technology are meaningless until the farmer actually uses the practices. The farmer's adoption or nonadoption of these practices, and the reasons underlying these behaviors, are critical dimensions for a comprehensive understanding of erosion and conservation processes.

Sociological contributions to soil conservation largely are responses to the questions of why accelerated soil erosion occurs, its impacts on farmers and communities, how to promote soil conservation, and who is adopting soil conservation and why. The last half century has seen a remarkable variety of responses to these questions, and perhaps of equal importance, a variety of methods by which answers to these questions have been pursued. The purpose of this chapter is to provide an overview of the sociological research that has examined these questions. Beyond a historical review of these studies, an explicit focus will be on assessing the extent to which the adoption process for soil conservation practices is predictable, and the research methods used in this effort. The intent is to learn from these past efforts in order to guide future research.

Sociology's Four Eras of Soil and Water Conservation Research

Rural sociologists have had an interest in soil erosion and soil and water conservation since the founding of their subdiscipline early this century. Over the years the specific substantive issues of interest within this area have shifted, as have the theoretical perspectives and research methods used. Although temporal divisions are always arbitrary to some degree, it seems that four fairly distinctive eras of sociological work in soil and water conservation during this century can be identified. The first is the Formative Era—1920s through 1940s, which coincides with the formative years of the rural sociology subdiscipline. The second era is the Organizational Era—1950s through mid-1970s, which is characterized by an emphasis on the organizations that oversee and deliver soil and water conservation programs. The third era is the Diffusion of Innovations Era— mid-1970s to late 1980s, so designated because of the predominance of the use of the diffusion of innovations model as the paradigm for explaining farmer use of soil and water conservation practices. The fourth era is loosely termed the Broadening Interests Era—late 1980s to present, because of the increasingly eclectic nature of the work being conducted in this era.

I. Formative Era—1920s through 1940s

Lowery Nelson (1937) captured sociological interest in this topic when he stated, "The conservation of the soil is not alone an economic and technological problem. In the last analysis it is a social concern." The Dust Bowl years of the 1930s certainly turned the attention of the nation to the necessity of natural resource conservation. Rural sociologists of this era, however, were interested in soil and water conservation even before the years of drought and devastating wind erosion. The main topic of interest was soil erosion depleting soil fertility, and the resulting challenges to the viability of farm operations and surrounding agricultural communities. In this era before the introduction of commercial nutrients and soil amendments, soil erosion was viewed as having an immediate and visible impact on viability of the farm operation. It was, as Nelson stated in the first volume of the new academic journal *Rural Sociology* in 1936, "Though true, it is not enough to say that soil erosion causes social disorganization." The social welfare of the farmer, family, and surrounding community were the focus of these early studies.

These early rural sociologists were not as interested in the causes of erosion or how to accelerate the adoption of soil conservation practices as became the norm in later years. Typical of discussions in rural sociology texts was Carl Taylor (1930, p. 12), "The loss of soil fertility, with the incident possibility of the destruction of the very foundation of farming, has recently become a problem of grave significance." Taylor continues, in somewhat less than scientific detachment, "The farmer can no longer mine the soil, he must husband and nurture it; and the realization of this fact has given him and the nation as a whole an attitude toward farming and its future, and toward the farmer's function which is different from any previous point of view" (Taylor, 1930, p. 12). In his rural sociology text, Charles Hoffer (1930) also discussed the problems of poor land:

> These are farmers marooned on some of this land with no hope of making enough to enjoy a reasonable standard of living. They have poor houses, a scanty food supply and a paucity of contacts that is extreme. It is almost impossible to develop a satisfying community life in such areas because the means of subsistence are meager (1930, p. 387).

Both Taylor and Hoffer did begin discussion of topics that became more prominent in the discipline some decades later—factors causing soil erosion and the promotion of soil conservation. Taylor (1930) cited tenancy as a major cause of soil erosion and by implication lack of soil conservation because tenants ". . . seldom find it to their advantage to follow any of these (soil conservation) practices . . . their tenure is usually too short to allow them to get the ultimate benefit" (1930, p. 250). Hoffer (1930, p. 385) stated, " . . . from the standpoint of the farmer it may be advantageous to get the greatest amount of profit from farming regardless of its ultimate effect on the fertility of the land." Anticipating future discussions of attitudes toward the land and the importance of an environmental ethic on the part of farmers for promoting soil conservation, Hoffer (1930, p. 386) suggested the necessity of developing " . . . on the part of farmers a pride in caring for their farms." This analysis of attitudes and opinions was also the theme of Russell Lord's book (1931), *Men of Earth*, that examined a variety of farming operations and practices.

The drought and dust bowl years of 1933-1935 focused attention on land degradation and particularly farmers' general lack of stewardship. In 1938 the U.S. Department of Agriculture's Yearbook of Agriculture had the subtitle *Soils and Men*, reflecting the importance of the relationship between biophysical and social processes. Taylor's input into this volume was prominent as lead author of a chapter on "The Distribution of People" in a section on "Public Purposes in Soil Use." The chapter carried on earlier rural sociology themes of maintaining viable farms and communities through moving farmers from the poorest lands, consolidating holdings to provide economically viable farm operations, and converting poor farmland into other uses. But in establishing these policies, Taylor et al. (1938, p. 56) stated, "The first concern is people, not soil, and especially people at the bottom of the economic pyramid."

In the latter part of this era, research on soil and water conservation and research on related topics with a significant component on soil and water conservation became more commonplace. The research methods included both the more traditional ethnographic studies of the time and also the social surveys rising in prominence in rural sociology. Among the more lasting works were the U.S. Department of Agriculture Bureau of Agricultural Economics' Rural Life Studies. In his forward to the reports of the six community studies, Taylor stated that they were chosen as points on a continuum from high community stability—a Lancaster County, Pennsylvania Amish Community—to great instability—a "Dust Bowl" Community in Kansas—with the other four communities somewhere between these extremes. The communities studied were El Cerrito, New Mexico; Sublette, Kansas; Irwin, Iowa; the Old Order Amish of Lancaster County Pennsylvania; Landoff, New Hampshire; and Harmony, Georgia.

As with earlier work, the researchers were interested in the impact of soil erosion on the viability of the farm operation and the community, and this is especially evident in the Sublette, Kansas (Bell, 1942) and Harmony, Georgia (Wynne, 1943) studies. But the studies also suggest that a more systematic effort was made to delineate factors that hinder or facilitate soil and water conservation practices. As some of the earlier literature suggested, tenancy and the immediacy of return on investment were found to be important in practicing conservation. Additional important factors were age (the younger, more progressive farmers were more likely to practice conservation), financial ability of the farmer, traditions related to plowing skills, and images of a good farmer which are contrary to contour plowing, peer pressure both for and against practicing conservation, and the physical characteristics of the land.

As already mentioned, concern about the viability of the farm and the larger community continued to be major issues with rural sociologists to the end of this era. Sanderson (1942) discussed both the factors related to soil erosion and soil conservation and the impacts of soil erosion in his volume *Rural Sociology and Rural Social Organization*, as did Hypes (1944) in his *Rural Sociology* article, "The Social Implications of Soil Erosion." Other articles appearing in *Rural Sociology* on this topic included Wakely (1936), "Social and Economic Effects of Soil Erosion," and Bui (1944), "The Land and the Rural Church." Stewardship and the role of the rural church in promoting stewardship were important themes during this era. Not only did the farmer have a moral obligation to practice stewardship, but there was a social obligation to the local community to be a good steward. Both the ethical and social norms regarding stewardship were advanced and supported in rural churches.

The research methods employed in this Formative Era were more diverse than in any other era with the possible exception of the current period, from the in-depth community studies using ethnomethodological techniques to informal observation and commentary. Innovative longitudinal approaches such as those by Cornell (1936) are similar to those employed today. Cornell studied the social, economic, and farm characteristics as well as farmer conservation behaviors prior to the implementation of a soil conservation project around Spenser, West Virginia. Plans included reexamining the area five years after the conservation demonstration project had commenced.* Hopkins and Goodsell (1937) used a questionnaire format to measure the conservation behavior of 400 Iowa farmers. Other than some discussion of the tenancy issue, there was little attention to other factors that may have explained the adoption of conservation techniques. A general practice was to compare the extent conservation was practiced on the farm to the overall viability of the farm operation. Descriptive reports of what constituted a "good farmer" always included an emphasis on conservation measures employed.

Perhaps the methodological diversity of this formative era is due to the singularity of the objective for sociological research. The welfare of farmers, their families, and rural communities were viewed as being directly related to conservation behaviors. Sociologists were interested in the adoption of conservation practices, not so much as a scientific exercise in prediction or explanation, but as a means of maintaining the viability of the rural landscape.

II. Organizational Era—1950s through mid-1970s

Although we have labeled the third era the diffusion of innovation's era, it is during the 1950s and 1960s that most diffusion of innovations research was accomplished in rural sociology and the model was developed and refined. In the early 1950s the North Central Rural Sociology Committee established the North Central Subcommittee for the Study of Diffusion of Farm Practices (Valende and Rogers, 1993). The most influential report of the subcommittee was How Farm People Accept New Ideas by George Beal and Joe Bohlen (1955). While the intellectual roots of the model were diverse and deep, this publication synthesized and crystallized the thinking on the diffusion of innovations. The model had its origin in the research of Bryce Ryan and Neal Gross (1943) on Iowa farmers' adoption of hybrid seed corn. Ryan and Gross were attempting to understand why this potentially profitable innovation required 13 years for Iowa farmers to fully adopt. Adoption behavior was viewed as the outcome of interaction among communication processes, characteristics of the innovation and the characteristics of the farmer. Farmers had access to different communication channels, moved through a specified decision process at different rates, and depending on the speed of this process,

*No evidence could be found if this "post-test" of the conservation treatment was ever completed. In all probability, and similar to contemporary conditions, there was little support for long-term social research.

were classified along a continuum from innovators at one end to laggards at the other. The typical research project examined the relationship of various factors such as personal or social characteristics of the farmer or farm family; values or attitudes of the farmer; characteristics of the farm operation; contacts with neighbors, dealers, or change agents; and sources of information on at least one but usually several improved farming practices or new farming technologies. Adoption and innovativeness were sometimes measured in terms of the use of individual technologies or practices, but more often as a scale created by combining several innovations. Usually the list of innovations would include one or more soil and water conservation practices such as strip cropping, contour plowing, or conservation tillage. Therefore the research was primarily a test of some aspect of the adoption and diffusion of an innovations model and only incidentally included variables important to the study of soil and water conservation. *Rural Sociology*, the primary journal in the subdiscipline, for all practical purposes was devoid of articles on soil and water conservation.

This lack of published research on soil and water conservation needs to be interpreted in the historical context of the period. The Great Depression was over in the early 1940s, but World War II held back the pending mechanical and biological revolutions in agriculture. Following World War II, and with the conversion of war industries to domestic production, we see an unparalleled unleashing of scientific and technological innovations that had been "waiting in the wings" the prior 10 to 15 years. Commercial nutrients, hybrid seeds, machinery innovations coupled with a generation of farmers who were better educated and more cosmopolitan in character led to the rapid adoption of many of these innovations. The productivity losses of excessive soil erosion could now be masked with commercial nutrients, and erosion was no longer viewed as the "national menace" of the earlier era. Faced with feeding the world as part of the reconstruction process, the agricultural experiment stations were now charged with understanding how to accelerate the adoption of these technical innovations. Consistent with the philosophy of the earlier era, it was felt that one bettered the conditions of farmers and rural communities by learning how to remove barriers to the adoption of these innovations.

In stride with the times, rural sociology textbooks included discussions of the Soil Conservation Service (SCS), the Agricultural Stabilization and Conservation Service (ASCS) and other related agencies, not to pursue the problems of soil erosion but to illustrate the operation of various USDA and other governmental agencies that were deemed relevant to farmers. Typical of such discussions were Rogers (1960) *Social Change in Rural Society* and Loomis and Beegle (1975) *A Strategy for Rural Change*. The discussions provided a historical and sociological analysis of the agencies with soil and water conservation being brought into the discussion only to illustrate what the agencies do and how they accomplish that work.

This organizational focus as the primary interest in soil and water conservation was reinforced by such *Rural Sociology* articles as, "Natural Leaders and the Administration of Soil Conservation Programs" (Hardin, 1957) and "Special Agency Program Accomplishment and Community Action Styles: the Case of Watershed Development" (Wilkinson, 1969). Hardin discussed the need for SCS to identify natural indigenous local leaders to assist in the successful implementation of their programs—an insider rather than an outsider approach. Wilkinson also addressed the problem of successfully implementing programs with a local community by an external state or federal agency, albeit through the use of concepts and actions suggested by community field theory.

The random sample defined the methodology in the research of this era. Even though earlier research had established a strong sociometric dimension (i.e., tracing communication patterns from one person or source to the next) to diffusion processes, and geographers had long studied the spatial diffusion patterns of innovations, a cross-sectional approach with a simple random sample of farmers was used in this era to study adoption behavior of farmers. More will be said about these methodological practices in the last section of this paper.

As a final note in closing this era, it is instructive to look at the first of a continuing series of volumes that examines contemporary issues, trends, and policy perspectives of rural America published under the auspices of the Rural Sociological Society. *Our Changing Rural Society: Perspectives and Trends* edited by James Copp and published in 1964 contained chapters on what were considered the important issues of the time and written by the most eminent of rural sociologists. There is no chapter in this volume on soil and water conservation, and as far as we can discern, no mention of soil conservation anywhere in the book. Obviously the scholars of that era did not deem soil and water conservation sufficiently important to include it in this prestigious book.

III. Diffusion of Innovations Era—mid-1970s to late 1980s

Just as the diffusion of innovation model was coming under increasing criticism, and research on the model was declining in rural sociology generally, it saw increasing use by rural sociologists working in soil and water conservation. The environmental movement had arrived by 1970 with widespread interest and support by the public for addressing environmental problems (Dunlap and Mertig, 1992). After 1975 a number of rural sociologists turned their attention to studying environmental and ecology related issues, and a popular model for conceptualizing this work, especially for farmers' use of soil and water conservation, was the diffusion of innovations model (Rogers, 1995). Although *Rural Sociology* published some of this research because of its applied nature, much more was published in other journals with the *Journal of Soil and Water Conservation* publishing the greatest share. The Soil and Water Conservation Society also published several books and special issues of the Journal during these years, all of which included significant contributions by rural sociologists.

With the diffusion of innovations model providing conceptual guidance and survey methodology the common data collection technique, the unit of analysis for research and the target for applied programs was the individual farmer. Much of the research focused on values, attitudes, and perceptions of the farmers in relation to concern with environmental issues, stewardship of the land, soil erosion, water pollution, and soil and water conservation practices. These social psychological traits were included with other variables relevant to the diffusion of innovations model to determine their efficacy in explaining the adoption of individual conservation practices or a summated scale of several practices. Other types of variables that often were included were the farmer's technological knowledge and skills; personal and family demographic and social characteristics; sources of farming information and use of the media; contact with change agents such as extension and the Soil Conservation Service; economic variables such as farm and nonfarm income, debt, and willingness to borrow money; farm operation characteristics such as number of acres owned and rented, ecological variables, such as erodibility of the land, types of commodities produced, and types of equipment and machinery used; and the perceived characteristics of the conservation practices such as relative advantage, cost, ease of use, and trialability. Additional variables sometimes included were political orientation and religious affiliation.

The whole range of farm conservation, from practices such as contour plowing to structures such as terraces was included in the research. Many articles focused on conservation tillage as a widely recommended and applicable conservation practice. Conservation tillage, however, is often loosely defined both by researchers and farmers. Some of the inconsistencies in results from research may be that farmers have inaccurate and widely varying perceptions of the practice and what it actually entails (Nowak and Korsching, 1985).

In this era, every aspect of the diffusion of innovations model was examined in the sociological research on soil and water conservation—innovation characteristics that affect adoption, along with the broad array of variables that affect the decision-making process. The decision-making process stages were examined, and the applicability of the adopter categories. One sociologist (Wagener, 1987) even examined the applicability of Cancian's (1979) upper middle class conservatism thesis for the adoption of innovations. And of course, the two or multistep flow of information and the role of opinion leaders in diffusion was included in the research.

Early in this era questions were raised about the diffusion of innovations model's applicability to environmental innovations. Pampel and van Es (1977) found that in a sample of Illinois farmers different variables predicted the adoption of commercial agricultural practices, and the adoption of environmental farming practices. They concluded that the model might not be relevant to noncommercial innovations. Their argument was countered first by Taylor and Miller (1978) that different farmers may have different orientations. Some farmers may have a more commercial orientation and others a more environmental orientation, and by incorporating this orientation the model remains useful in predicting both types of innovations. Further debate in the literature occurred between van Es (1983) and Nowak (1983, 1987). Nowak makes the point that conservation practices are not necessarily unprofitable and that there are both micro and macro factors that influence adoption.

By this time the macro factors already had begun to be incorporated into diffusion of soil and water conservation research. It was given impetus through the publication of Lawrence Brown's volume, *Innovation Diffusion* in 1981. In this volume, Brown, a geographer, outlined the market and infrastructure model. Whereas the classical diffusion of innovations model is largely a social psychological communication and decision-making model—a demand model, Brown's model outlines the necessary infrastructure and strategies for delivering the innovations—a supply model. Toward the end of this era increasingly more research included what might be called market and infrastructure variables such as educational programs, technical and financial assistance, interagency cooperation in delivering programs, and other types of institutional support or deterrents.

Through the mid-1980s the primary approach to soil and water conservation was the voluntary approach, and institutional support in the form of educational, technical, and/or financial assistance is especially important for the voluntary approaches. In one of the more ambitious efforts to evaluate the voluntary approach, Nowak (1982) implemented a pilot program in a 16-county southeast Iowa area using the two diffusion of innovations models. After conducting a pretest to determine farmers' needs for practicing good soil and water conservation, two distinct educational programs were developed—one each for the supply and demand models. Each program was randomly assigned to four counties, four counties received both programs, and four counties were assigned as controls receiving no programs. Unfortunately, as with so many federally supported programs, funding was terminated prematurely. Evaluation results from the follow-up surveys showed little impact, in large measure owing to the early termination and insufficient time for the program to have an impact (Korsching et al., 1985; Hoban et al., 1986). Even so, the educational materials that were developed for this program have been modified, refined, and published for broader circulation and application (Hoban 1992a, 1992b, 1994).

One additional issue seems worthy of comment at this juncture. A consistency in the results of many research projects was what has been termed the "proximity effect." When farmers were asked either whether there was an erosion problem, or the seriousness of the erosion problem, the closer the point of reference to the farmer's own operation, the less serious the perceived problem. As one moves away from the farm operation, the more serious the problem is perceived to be. Although often reported and discussed as an interesting finding, to our knowledge these results were never formally published in a journal article or scholarly book with a theoretical rationale for the finding.

IV. Broadening Interests Era—Late 1980s to Present

Much of the sociological soil and water conservation literature in the Broadening Interests era relates to the topics of soil conservation, water quality, resource management policy, and sustainable agriculture. Some of the work on the first issue, conservation, was a continuation of earlier concerns with voluntary versus mandatory approaches spurred on by the new federal cross compliance policies. Through cross compliance regulations, farmers' eligibility for various federal farm program benefits was contingent initially upon having a conservation plan for the farm by 1990 and later having that plan implemented by 1995. The shift in research efforts was from understanding and predicting the adoption process to understanding and predicting the extent that institutional arrangements influenced adoption processes. Although traditional incentives in the form of technical and financial assistance continued to be available to farmers, the more compulsory strategies break a longstanding tradition and norm of the farmer having control of land use on the farm. Sociologists (Napier, 1994) evaluated the effectiveness and impacts of the compulsory strategies. Sociologists also were interested in the implementation and impacts of the Conservation Reserve Program, especially landowners' decisions and actions regarding CRP land upon program termination (Nowak et al., 1991). An additional policy instrument for water quality examined by sociologists was micro-targeting of specific areas within a watershed based on environmental degradation, as opposed to general programs based on gross erosion rates (Lovejoy et al., 1985).

Another major topic of broadening interests, sustainable agriculture, was a logical continuation of early interests in soil and water conservation where the concern was on the viability of the farm family and surrounding rural community. Along with farm profitability and food security, soil and water conservation broadly defined are primary dimensions of sustainable agriculture. Sociologists are providing strong leadership in examining, defining, and promoting sustainable agriculture. Iowa State University's Leopold Center for Sustainable Agriculture has a number of issue teams in the soil, plant, and animal sciences—each including a sociologist, as well as a human issues team to which sociologists provide primary leadership.

Sociologists have addressed a number of topics in their work that fall under the rubric of sustainable agriculture. These topics include the adoption of integrated pest management practices (Thomas et al., 1990), farm operator perspectives and practices on the environmental consequences of animal waste disposal (Molnar and Wu, 1989), issues related to farmers' participation in wetlands programs (Lovejoy, 1991), factors related to reduction in use of commercial fertilizers and chemicals by farmers (Lasley et al., 1990; Korsching and Malia, 1991), and factors related to farmers' adoption of new sustainable agriculture technologies such as the late spring soil nitrate test for corn (Contant et al., 1991).

Concomitant with the substantive broadening is a broadening in theoretical perspectives and methodological approaches to soil and water conservation issues. The diffusion of innovations theoretical model which was so predominant in the work of the previous era is less often and sometimes not at all cited in this era's literature. In an effort to better understand the conservation, or lack of conservation behavior of farmers, different theoretical perspectives are being applied such as exchange theory (Napier and Napier, 1991) and social learning theory (Napier and Brown, 1993). Moving away from the diffusion of innovations model also has encouraged units of analysis other than the individual farmers in research and program development. This trend, originating in the previous era with the use of the market and infrastructure model, includes the role and capacity of conservation districts in resource management (Nowak, 1985), organizational and institutional factors leading to failure of externally imposed natural resource programs (Malia and Korsching, forthcoming), agribusiness manager perceptions of new sustainable agriculture technologies, specifically, the late spring soil nitrate test for corn (Korsching et al., 1992), the role of agrichemical dealers in providing analytical and diagnostic services that reduce reliance on purchased inputs (Wolf and Nowak, 1995) and the nature of general public support for conservation programs (Molnar and Duffy, 1988).

In this discussion of broadening interests we must be cognizant of the fact that the broadening occurred not only substantively and conceptually, but also geographically. The sociological literature on soil and water conservation increasingly addresses issues beyond the borders of the U.S. Although an occasional work (article, chapter, thesis, or dissertation) by a sociologist outside the U.S. appeared in the earlier years of the last two decades, only recently has there been anything that seemed to approach a critical mass. Such works as the recently published *Adopting Conservation on the Farm: An International Perceptive on the Socioeconomics of Soil and Water Conservation* (Napier et al., 1995) will go a long way toward the development of an international sociological body of knowledge on soil and water conservation from which all contributors should benefit.

Methodological Issues

Beyond this cursory review, there is a profuse research tradition of examining the development, characterization, and control of new agricultural conservation technologies; the individual and farm firm factors that influence decision processes surrounding the use of these methods; the role of institutions and markets in governing availability and access to these innovations; the communication and institutional factors influencing the dispersion of these technologies across space and time; and to a certain extent, the impact or consequences of these innovations on individuals, social groups, and the environment (Arnon, 1989; Clark, 1987; Glaser et al., 1983; Rogers, 1995).

In disregard of this broad and impressive research tradition, several scholars have concluded that the use of adoption models relative to natural resource management practices are inappropriate, ineffectual, and perhaps even irrelevant (Goss, 1979; Napier, 1989; Pampel and van Es, 1977). Others have called this research tradition "atheoretical" (Hooks, 1980). Schaller (1992, p. 254), reviewing a book on ecological agriculture, noted that he was "struck by the enormity of the barriers to adoption of farming in nature's image and dismayed by the lack of social science research—equal in imagination and rigor to the authors." Nowak (1987) noted that "much of the research in this area has been poorly conceived and rarely moved beyond perceptual data and gross measurements." Even back in 1968 during the apogee of adoption and diffusion research, Coughnenour (1968, p. 6) noted that significant detail is being ignored or glossed over in the overly simplified approach employed, and that "the entire area of adoption is badly in need of reformulation."

In 1990 William Lockeretz published an article in the *Journal of Soil and Water Conservation* entitled, "What Have We Learned About Who Conserves Soil?" After reviewing a substantial body of research literature on the topic, most of it published by economists and sociologists, Lockeretz came to the conclusion that the literature provides little information about which farmers conserve soil and why. Pointing out serious methodological, conceptual, and statistical flaws characterizing this body of research, he concludes by noting that we need to rethink the basic approach used in past research. In sum, while there has been a significant amount of research on the adoption of various natural resource management practices, continuing and pervasive criticisms call into question the utility of this whole line of inquiry. The purpose of this section of the chapter is to examine the methodological and conceptual features of this body of research while suggesting possible future directions. However, rather than taking the direction of Lockeretz (1990), who called for a completely new approach, it is our belief that there is significant utility in the adoption and diffusion of innovations model if certain conceptual and methodological issues are addressed.

Issues Associated with Research Design and Concept Measurement

The simple random sample has been the mainstay sampling procedure of sociologists studying various rural issues. The principal assumption has been that the phenomenon being studied, be it a conservation behavior, a stewardship attitude, or an erosion belief, is randomly distributed in the rural population. A research design built around a simple random sample, therefore, will adequately "capture" this phenomenon being studied. This assumption, either explicitly or implicitly, has been carried over into the study of the adoption of natural resource management practices. Many of the adoption studies reviewed by Lockeretz and other critics have been based on random sampling techniques.

Yet forms of natural resource degradation are not randomly distributed across the landscape. Degradation (e.g., accelerated erosion, water quality pollution) is the outcome of an interaction between human behaviors (e.g., tillage or agrichemical management practices) and the site-specific features of the biophysical environment where these practices are employed. Even a cursory examination of any depiction of cropland susceptibility to erosion or vulnerability to surface or groundwater pollution shows specific spatial patterns (for examples see Kellogg et al., 1994 relative to leaching of pesticide and nitrate and Hamlett et al., 1992 for nonpoint surface water pollution). Combine these biophysical spatial patterns with the spatial diffusion pattern of innovations that have been the basis of geographic research (Brown, 1981), and one must question the validity of a sample based solely on random techniques. If the spatial pattern of accelerated erosion—the result of the interaction between biophysical features and agronomic behaviors—is not randomly distributed across a landscape, then why should the sample frame for measuring conservation behaviors be based on random techniques? Further, assuming some logic to the design of government programs and implementation efforts (Heimlich, 1994), one would expect the probability for conservation behaviors to be greatest in those areas with the greatest need (i.e, targeting). All this implies that the sample frame needs to be constructed with an understanding of the phenomenon being studied. Traditional conservation practices such as terraces, diversions, and grassed waterways would most likely be promoted, and to a certain extent adopted, in areas with large LS (length of slope and steepness of slope) coefficients from the Universal Soil Loss Equation (USLE). Proportionate spatial sampling on this feature would allow a more valid test of adoption of traditional conservation practices.

An issue clearly related to the development of a valid sampling frame is including a control for the applicability, need, or appropriateness of the conservation practice. All too often researchers have simply constructed a list of general conservation practices, asked a sample of farmers which practices on the list they are using, and then classified the farmer on adoption behavior or innovativeness based on the summated scale. Little effort is made to control on the appropriateness of individual practices relative to the biophysical features of the farm. It is as Ashby (1982, p. 235) noted:

> The suitability of agricultural technologies to different farming environments is addressed only indirectly in the diffusion literature in terms of the availability of socio-economic resources which facilitate or inhibit farmers' innovation, while the physical and natural parameters of agriculture are largely ignored.

Farmers having land with little need for a practice are treated the same in the analysis as farms having great need, and if neither farmer adopts, their behavior is considered equivalent despite the appropriateness of the behavior for one of the farmers but not the other. Needless to say, this error introduces significant "noise" into the model results. Very few of the studies examining the adoption of natural resource management practices have included controls for conservation needs (see Erwin and Erwin, 1982 for an example of an exception to this observation). At best, the study is "weighted" or "stratified" to give preponderance to areas where there is greatest need for the remedial practices being examined (see Duff et al., 1992; Napier et al., 1988). At worst, no consideration is given to biophysical features in the construction of the sampling frame.

Several issues relate to precisely what constitutes adoption of conservation practices. The primary goal of research is to explain adoption behavior. That is, most studies attempted to predict the adoption of various conservation measures using a long list of personal, farm firm, communication, market and infrastructure, and other predictor variables. Yet there was little consistency across these studies in how this concept of conservation behavior was measured. As simple as it may sound, the question "What is conservation behavior?" has had a variety of responses. It is difficult if not impossible to develop valid and reliable models of conservation behavior without some agreement on the measurement of the dependent variable. This diversity can be seen in Table 1 where the focus is on a single practice, conservation tillage. Conservation behavior in general, and conservation tillage in particular, have been measured by practices used (both claimed and based on actual measurements), financial investment in conservation practices, changes in erosion rates across time, and current erosion rates. These latter erosion adoption measures have been applied to a specific field whereas the practice use and investment measurements have been applied to the farm as a whole.

While financial investment does capture one dimension of proportionate effort on the part of the adopter, it is less than ideal on two others. First, investment does not capture the environmental impact of that investment. For example, the investment in a new tillage system and a traditional erosion control structure may be similar, but the impact of each investment on erosion rates could differ dramatically (e.g., the "soil saved" with 200 feet of terrace versus that saved with 30% residue across all cropland on the farm). Second, financial investment does not capture the complexity associated with practices heavily dependent on managerial expertise. Some of the residue management tillage systems and nutrient and pest management practices are complex, yet require little financial investment. One resolution to this problem of measuring adoption behavior is to view it as a "proportion of needed effort" relative to resource management. This could be the result of examining the set of erosion rates, soil tolerance values and conservation behaviors, or it could be reduction in loading to surface or groundwaters based on adopting certain practices. Regardless of the acceptability of this alternative, more effort needs to be given to measuring the concept of conservation behavior.

Another problem is the level of measurement applied to adoption behavior. Some studies have treated adoption as a dichotomous measure (a farmer is either a adopter or a nonadopter). Other studies have treated it as an interval (e.g., high, medium, and low level of adoption) or ratio variable (percent of

Table 1. Conceptual Dimensions of Selected Conservation Tillage Adoption Studies

Author(s) of Publication	Conservation Practice(s) Examined	Measurement of Adoption Behavior	Account for Practice Suitability to Physical Setting	Account for Accuracy or Extent of Adoption
Abd–Ella et al., 1981	Conservation tillage	Description of primary tillage tool	Not considered	Not considered
Allmaras & Dowdy, 1985	Conservation tillage	USDA survey of tillage	Climate and major cropping systems	Not considered
Belknap & Saupe, 1988	Conservation tillage	Claimed use of any tillage other than moldboard plow	Not considered	Not considered
Dickey et al., 1986	Conservation tillage	Claimed use & residue measurements	Not considered	Yes, 3 transects at random places on field
Green & Heffernan, 1986	Conservation practices & tillage	Claimed use on survey	Not considered	Not considered
Ladewig & Garibay, 1983	Conservation tillage	Claimed use	Used cropping regions, erosion rates and crops	Not considered
Napier et al., 1988	15 conservation practices including tillage	Claimed frequency of use	Not considered	Not considered
Norris & Batie, 1987	Conservation practices including tillage	Annual investment in practices	Not considered	Not considered
Nowak & Korsching, 1985	Conservation tillage	Claimed use versus residue levels	Not considered	Measured on field basis

practices adopted from a list of practices). The literature clearly has shown that adoption is a process, not a discrete event. The individual goes through a number of stages (e.g., awareness of the innovation, evaluation, first use, and adaptation) in the decision process with rejection a possibility at each stage. Collapsing this process into a dichotomous event could be partially responsible for Lockeretz's (1990) conclusions about the poor predictive record of research using this model.

Table 1 also illustrates two other dimensions of adoption studies that have been rarely considered in past work. These are the extent the practice is employed on applicable acres, and the accuracy with which the practice is being used. A basic step for any conservation technician is to examine the severity and distribution of erosion on a farm to determine the nature and extent that different conservation practices need to be employed. Nonetheless, the extent that a practice is employed on applicable situations is rarely considered in the adoption research literature. Individuals are treated as adopters of a practice regardless of whether they apply the practice on 10%, 50% or 100% of the situations requiring this practice. In their criticism of adoption research, Downs and Mohr (1976) state that to capture the variations in behavior we really want to explain, we must measure the extent of implementation. A simple adoption/nonadoption operationalization can result in gross measurement error and introduce additional "noise" in the application of the adoption model in natural resource management. Moreover, individuals who may have tried a practice and rejected it are treated as nonadopters with no attempt to learn why this decision was deemed necessary.

Accuracy in the use of a practice when it is adopted is another dimension ignored in past research, yet this will become more important in the future with increasing sequential interdependence and complexity of practices applied in the field. Using a conservation practice and using this practice so that it achieves its intended function are not necessarily the same. Adoption of a practice or technology is not equivalent to conservation behavior. Work by Nowak and Korsching (1985) and Dickey et al. (1986) established this fact rather conclusively relative to conservation tillage. The studies found that farmers were defining the use or adoption of conservation tillage based on the primary tillage tool—not how that tillage tool was used relative to residue management. It is not enough to simply ask farmers whether or not a practice is being used. One also needs to measure how it is being used to assess if intended functions are being achieved. Bohlen (1968, p. 20) recognized this fact when he stated:

> . . . many ideas get rejected, not because the idea wasn't good, but because the person carrying out the trial didn't follow directions or in some other way failed to use the idea as recommended.

All of the methodological comments are not meant to criticize past research. Our comments are meant to point out the tremendous opportunities for future research in the adoption of soil conservation practices and behavior. As Lockeretz (1990, p. 523) states, "It is time to step back and carefully look over what has been done so that we can derive more value from future efforts."

The Future of Adoption Research: Capturing Time and Space

Adoption and diffusion of conservation practices occur across time and space. Our research designs and efforts must be able to capture this fundamental fact. The richness and diversity of the biophysical environment that farmers manipulate on a daily basis is a critical factor in assessing the appropriateness of remedial practices, the extent these practices can be used as designed or need to be adapted, and the consequences of practice use. Layered over this diverse biophysical setting is the institutional environment. It also is very diverse in nature contrary to the image of uniform national policies applied homogeneously across the landscape. This diversity emerges from three sources: (1) state policies that may augment or detract from the objectives of national policy; (2) variation in implementation of national policy due to varying fiscal resources, capability of local program managers, and the degree of ambiguity in applying national administrative rules; and (3) the strength and congruity of private sector markets in supporting, subverting, or ignoring recommendations based on national policy (Wolf and Nowak, 1995). Finally, on top of biophysical and institutional diversity we superimpose the diversity of farm operators and their social organizations—the traditional focus of adoption studies. The adoption and diffusion of soil and water conservation practices occurs within this diverse, multilayered context across time and space. Many of the deficiencies of past research can be linked to sample frames unable to capture this diversity. A sampling technique is needed that can link "people" with "place" across time. As Lockeretz (1990, p. 523) states:

> . . . studies of the kind described here should be done over a wide enough area and long enough period that we can learn how variable or stable the results are with respect to time and location. Unfortunately, such studies would require considerably more money than is typically available.

This needed perspective is seen in Figure 1. While time is not graphically represented, it is implicitly "woven" into any research design employing this perspective due to the dynamic nature of the layers. At issue, then, is how to create this sampling frame that can capture both people and place.

Yates (1981) and Haining (1990) outline a rich history of thought regarding sampling frames in the social and environmental sciences. Yates maintains that a good sample should: (a) minimize selection bias due to errors of judgement and (b) minimize errors due to chance inclusion or exclusion of population members into the sample. Adoption studies have traditionally attempted to avoid these biases by using some form of a random sample of individuals to

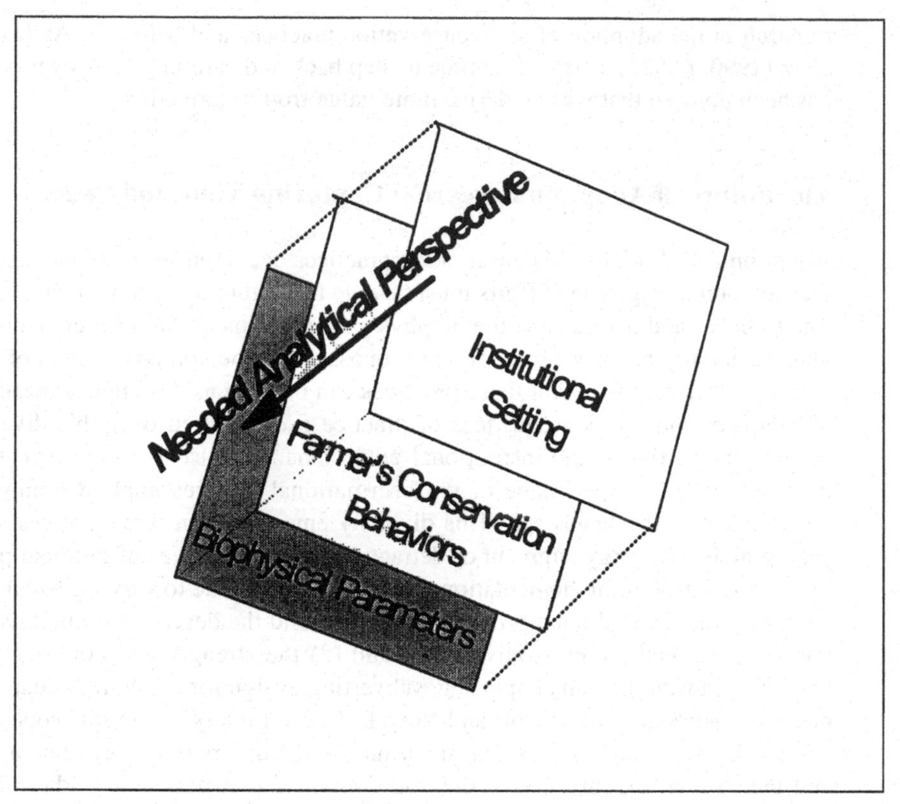

Figure 1. "People" and "Place" Perspective.

whom inferential statistics are applied. Yet, as noted earlier, a random sample of individuals does not address the applicability of the conservation practices being studied; i.e., it does not capture "place" where this process is occurring. The adoption of conservation practices must also sample on, or at least control for, the land on which these practices are to be applied. Geographers and other earth scientists have much to offer to the social sciences in this area of study.

Sampling across the land may be random, stratified, systematic, or combinations of these (Clark and Hosking 1986; Dalton et al., 1975). When sampling land surfaces, a simple random sampling design (with replacement) meets a main tenet of inferential statistics by providing equal probability that each "site" on the surface could be selected. Yet, random sampling may not give a thorough coverage of the sample area, leaving spatial gaps (Haggett et al., 1977; Wrigley 1979). As for list-generated samples, random geographic selection of sites may also be clumped, causing bias due to positive spatial autocorrelation; i.e., closely-spaced items resembling one another more closely than items which are farther apart (Williams, 1984). A systematic sample across space addresses this deficiency in that it selects sites for inclusion at regular intervals (Matérn, 1960). While this provides more complete coverage of an area, the regular spacing of the sample may cause a bias by over- or underrepresenting periodic phenomena (Griffith, 1987; Barber, 1988).

Location sampling designs have evolved that combine the positive aspects of both random and systematic techniques, while minimizing the negatives (Holmes, 1967; Haggett, 1990). Berry and Baker (1968) developed a technique that benefits from the spatially unaligned aspect of random sampling, while providing the blanket coverage of a systematic sample. The general design for this stratified systematic unaligned sampling had been used earlier by Berry (1962) for studying agricultural land within flood plains. Uses of unaligned systematic techniques, such as geologic applications cited by Krumbein and Graybill (1965) and Berry's studies have focused on land cover or physical attribute inventories. The USDA also uses a stratified systematic unaligned framework to choose the sample sites for their periodic National Resource Inventory.

One links "people" and "place" in examining the adoption of conservation practices by developing a sampling frame that is: (1) stratified or weighted toward those biophysical conditions where there is a high probability the innovation (conservation practice) would be deemed applicable; (2) systematic in that the sampling frame selects sites that cut across salient institutional and farm operator diversity; and (3) unaligned so as to proportionately capture biophysical irregularities (e.g., glacial moraines or sinkholes in karst topography) or spatial diffusion patterns (e.g., intense private sector competition driving adoption rates in some areas but not others). This type of research design will address Lockeretz's concern on the ability to aggregate studies across place, and it will lead to the standardization of measures due to the explicit incorporation of biophysical features.

Developing an appropriate sampling frame will not, by itself, address previously discussed deficiencies or strengths of adoption research. Measurement of critical concepts must also be addressed. Yet the very fact that "place" is an explicit component of the sampling frame should increase the validity and reliability of these measures. When one specifically accounts for the biophysical setting where the innovation is used, the extent and accuracy to which this innovation is used is a logical extension. Further, issues of precisely defining conservation behavior (e.g., practice use, changes in degradation rates, investment in conservation, etc.) that have plagued past research are now conditioned by knowledge of the biophysical setting. The setting rather than some rigid decision rule defines adoption behavior. In this approach conservation behavior could be a proportionate measure of needed effort as suggested earlier, or it could simply be measuring the behavior in a fashion that is conducive to replication in other studies.

Summary and Conclusion

Accounting for spatial settings and patterns of conservation behavior across space and time will largely determine the power and utility of future adoption studies. Including these dimensions will provide an opportunity to determine the degree of generalizability of research results and applied programs, and perhaps provide new insights into issues and problems eluding definitive answers or practical solutions in the past.

Although predicting future directions for soil and water conservation work by sociologists is difficult, another avenue that holds much promise is the incorporation of indigenous knowledge. Incorporation of indigenous knowledge is a strategy first used in international development. Earlier in the adoption of innovations literature farmers were given credit for reinventing technology as they adopted and implemented it, but their own knowledge and technologies often were discounted in the face of "scientifically" developed knowledge and technologies. McCallister's work (1996) on how farmers understand their soil found significant differences between farmers and formal characterizations (i.e., modern county soil surveys) of this same soil resource. Understanding how farmers understand their soil could be an important first step of bringing indigenous knowledge into the study of conservation behavior. The contributions indigenous knowledge can make to agricultural development generally and soil and water conservation specifically is now being recognized by sociologists (Dialla, 1993).

Whether the focus is on scientific or indigenous knowledge, sociologists will continue to be involved in the dialogue on environmental policies and programs for agriculture; in research to define the issues, problems, and solutions to soil quality degradation and water pollution; and in the implementation of the research results to achieve sustainable soil and water conservation. Over the last 50 years, progress has been made in demonstrating the importance of social factors in achieving soil and water conservation, but the knowledge base remains limited. No listing of empirical generalizations will be attempted here. That is another chapter (or volume) requiring a much more comprehensive review of the literature than just completed.

In soil and water conservation, sociology's biggest battle has been to achieve recognition from agronomists, engineers, soil scientists, and economists, among others, of the role of nonecological, nontechnological, and noneconomic factors as making important contributions in explaining farmers' behavior. In this quest sociologists have achieved a reasonable amount of success. This "battle" for legitimacy may have led sociologists to focus undue attention on the sociological while underemphasizing the biophysical and institutional context of these social processes. The lesson is perhaps best captured in the anomaly of the "proximity effect," in which the perceived erosion problem may have no relationship to the amount of erosion actually occurring on the farmer's field in question. Yet the farmer's conservation behavior will be based on a fundamental sociological principle—what is perceived as real is real in its consequences. The same principle may be applied to sociological contributions to soil and water conservation over the last 50 years. Regardless of how this contribution is perceived by fellow researchers, policy administrators, and farmers, the real issue is the perception sociologists will develop in the next 50 years. To the extent that we are cognizant of the significant contributions of past research, yet aware of the opportunities to integrate space and time into future research, then it is our perception that the main sociological contributions to soil and water conservation are yet to come.

References

Abd-Ella, M., E. Hoiberg, and R. Warren. 1981 Adoption behavior in family farm systems: An Iowa study, Rural Sociol., 46(1):42–61.

Allmaras, R., and R. Dowdy. 1985. Conservation tillage systems and their adoption in the United States. J. Soil Water Conserv. 5:197–222.

Arnon, I. 1989. Agricultural Research and Technology Transfer. Elsevier Applied Science, NY.

Ashby, J. 1982. Technology and ecology: Implications for innovation research in peasant agriculture. Rural Sociol. 47 (2):234–250.

Barber, G. 1988. Elementary Statistics for Geographers. Guilford Press, N.Y.

Beal, G. and J. Bohlen. 1955. How farm people accept new ideas. Cooperative Extension Report No. 15, Iowa State University, Ames, IA.

Belknap, J. and W. Saupe. 1988. Farm family resources and the application of no–plow tillage in south western Wisconsin. North Central J. Agr. Econ. 10(1):14–23.

Bell, E. 1942. Culture of a contemporary rural community: Sublette, Kansas. Rural life Studies: 2, Bureau of Agricultural Economics, U.S. Department of Agriculture, Washington, D.C.

Berry, B. 1962. Sampling, coding and storing flood plain data. Agric. Handbook No. 237, USDA, USGPO. Washington D.C.

Berry B. and A. Baker. 1968. Geographical sampling. pp. 135–154 In: Spatial Analysis: A Reader in Statistical Geography. Prentice-Hall, Englewood, NJ.

Bohlen, J. 1968. Research needed on adoption models. pp. 15–21 In: Diffusion research needs. North Central Regional Research Bulletin 186: University of Missouri Agricultural Experiment Station, SR89 1-68-2.5M. Columbia, MO.

Bromely, D. 1991. Environment and Economy: Property Rights and Public Policy. Basil Blackwell, Inc., Cambridge, MA.

Brown, L. 1981. Innovation Diffusion: A New Perspective. Methuen, New York.

Buie, T. 1944. The land and the rural church. Rural Sociol. 9(3):251–257.

Cancian, F. 1979. The Innovators Situation. Stanford University Press, Stanford, CA.

Clark, P. 1987. Anglo American Innovation. Walter de Gruyter, Berlin.

Clark, W. and P. Hosking. 1986. Statistical Methods for Geographers. John Wiley & Sons, NY.

Contant, C., P. Korsching and C. Young. 1991. Farmers' assessment of the late spring soil nitrogen test (N-Trak). Paper presented at the Annual Meeting of the International Society of Impact Assessment, Champaign, IL.

Copp, J. 1964. Our Changing Rural Society: Perspectives and Trends. Iowa State University Press, Ames, IA.

Cornell, F. 1936. A social and economic survey of the Spenser soil–conservation area. West Viginia Agricultural Experiment Bulletin 269.

Coughnenour, M. 1968. Some general problems in diffusion: From the perspective of the theory of social action. pp.. 5–14 In: Diffusion research needs. North Central Regional Research Bulletin 186: University of Missouri Agricultural Experiment Station, SR89 1-68-2.5M. Columbia, MO.

Dalton, R., J. Garlick, R. Minshull and A. Robinson. 1975. Sampling Techniques in Geography. George Phillip and Son, Ltd., London.

Dialla, B. 1993. The adoption of soil conservation practices in Burkina Faso: The role of indigenous knowledge, social structure and institutional support. Unpublished Ph.D. dissertation. Department of Sociology, Iowa State University, Ames, IA.

Dickey, E., P. Jasa, B. Dolesh, L. Brown and K. Rockwell. 1986. Conservation tillage: Perceived and actual use. J. Soil Water Conserv. 42(6): 431–434.

Downs, G. And L. Mohr. 1976. Conceptual issues in the study of innovations. Adm. Sci. Quart., 21 (4):700–714.

Duff, S., D. Stonehouse, D. Blackburn, and S. Iltilts. 1992. A framework for targeting soil conservation policy. J. Rural Studies 8(4):399–410.

Dunlap, R. and A. Mertig. 1992. American Environmentalism: The U.S. environmental movement, 1970–1990. Taylor and Francis, Philadelphia, PA.

Erwin, C. and D. Erwin. 1982. Factors affecting the use of soil conservation practices: Hypotheses, evidence and policy implications. Land Econ. 58(3): 277–292.

Glaser, E., H. Abelson, and K. Garrison. 1983. Putting Knowledge to Use. Jossey–Base Publishers, San Francisco.

Goss, K. 1979. Consequences of diffusion of innovations. Rural Sociol. 44(winter):754–772.

Green, G. and W. Heffernan. 1986. Government programs for soil conservation: progressive or regressive effects. Environ. Behav. 18(3):369–384.

Griffith, D. 1987. Spatial Autocorrelation: A Primer. Association of American Geographers, Washington D.C.

Haggett, P., A. Cliff, and A. Frey. 1977. Locational Analysis in Human Geography. Edward Arnold Ltd., London.

Haggett, P. 1990. The Geographer's Art. Basil Blackwell, Ltd., Oxford, England.

Haining, R. 1990. Spatial Data Analysis in the Social and Environmental Sciences. Cambridge University Press, Cambridge, England.

Hamlett, J., D. Miller, R. Day, G. Peterson, G. Baumer, and J. Russo. 1992. Statewide GIS-based ranking of watersheds for agricultural pollution prevention. J. Soil Water Conserv. 47(5) 399–404.

Hardin, C. 1957. 'Natural leaders' and the administration of soil conservation programs. Rural Sociol. 16(3):279–281.

Heimlich, R. 1994. Targeting green support payments: The geographic interface between agriculture and the environment. pp. 11–49 In: Sarah Lynch (ed) Designing Green Support Programs. Report # 4 Wallace Institute Policy Studies Program, Washington, D.C.

Hoban, T. 1994. Managing conflict. Conservation Technology Information Center, West Lafayette, IN.

Hoban, T. 1992a. Teamwork for conservation education. J. Soil Water Conserv. 47(3):231–233.

Hoban, T. 1992b. Strategies for building a conservation team. J. Soil Water Conserv. 47(4):294–297.

Hoban, T. P. Korsching, and T. Huffman. 1985. The Selling of Soil Conservation: A Test of the Voluntary Approach, Vol. 2: Organization Survey. Sociology Report 158, Department of Sociology and Anthropology, Iowa State University, Ames.

Hoffer, C. 1930. Introduction to Rural Sociology. Richard R. Smith, Inc., New York.

Holmes, J. 1967. Problems in locational sampling. Ann. Assoc. Am. Geog. 57:757–780.

Hooks, G. 1980. The classical diffusion paradigm in crisis. Paper presented at the Annual Meeting of the Rural Sociological Society, Ithaca, N.Y.

Hopkins, J. and W. Goodsell. 1937. Practices on Iowa farms. Iowa Agr. Expt. Sta. Bull. 360.

Hypes, J. 1944. The social implications of soil erosion. Rural Sociol. 9(4):363–375.

Kellogg, R., M. Maizel, and D. Goss. 1994. The potential for leaching of agrichemicals used in crop production: A national perspective. J. Soil Water Conserv. 49 (3):294–298.

Korsching, P., C. Contant, S. Wilson, and C. Young. 1992. Fertilization dealers' perceptions of the impacts of the late spring soil nitrate test on business viability. Impact Assessment Bull. 10:71–85.

Korsching, P., T. Hoban, and J. Maestro–Scherer. 1985. The selling of soil conservation: A test of the voluntary approach, Vol. 1: Farmer Survey. Sociology Report 157, Department of Sociology and Anthropology, Iowa State University, Ames.

Korsching, P. and J. Malia. 1991. Institutional support for practicing sustainable agriculture. Am. J. Alternative Agric. 6:17–22.

Krumbein, W. and F. Graybill. 1965. An Introduction to the Statistical Models in Geology. McGraw-Hill, NY.

Ladewig, H. and R. Garibay. 1983. Reasons why Ohio farmers decide for or against conservation tillage. J. Soil Water Conserv. 38(6):487–488.

Lasley, P. M. Duffy, K. Kettner, and C. Chase. 1990. Factors affecting farmers' use of practices to reduce commercial fertilizers and pesticides. J. Soil Water Conserv. 45(1):132–136.

Lockeretz, W. 1990. What have we learned about who conserves soil? J. Soil Water Conserv. 45(5):517–523.

Loomis, C. and J. Beegle. 1975. A Strategy for Rural Change. Schenkman, Cambridge, MA.

Lord, R. 1931. Men of Earth. Longmans & Green and Co., New York.

Lovejoy, S. 1991. Wetlands: Sell to the highest bidder? J. Soil Water Conserv. 46(6):418–419.

Lovejoy, S., J. Lee, and D. Beasley. 1985. Muddy water and American agriculture: how to best control sedimentation from agricultural land. Water Resour. Res. 21(8):1065–1068.

Malia, J. and P. Korsching. 1996. Local vs. societal interests: Resolving a community development dilemma. J. Community Dev. Soc. Forthcoming.

Matérn, B. 1960. Spatial variation: Stochastic models and their application to some problems in forest surveys and other sampling investigations. Meddelanden Fran Statens Skogsforsloninginstitut 19:1–144.

McCallister, R. 1996. How Wisconsin farmers understand and manage their soils: A site-specific people and place methodological analysis. Unpublished Ph.D. dissertation, Institute for Environmental Studies, Land Resources Program, University of Wisconsin-Madison.

Molnar, J. and L. Wu. 1989. Environmental consequences of animal waste disposal: Farm operators perspectives and practices. Circular 297, Alabama Agricultural Experiment Station, Auburn University, Auburn, Alabama.

Molnar, J. and P. Duffy. 1988. Public perceptions of how farmers treat soil. J. Soil Water Conserv. 43(2):182–185.

Napier, T. 1989. Implementation of soil conservation practices: past efforts and future prospects. Topics in Appl. Resour. Management Tropics 1:9–34.

Napier, T. 1994. Regulatory approaches to soil and water conservation. pp. 189–202 In: L. Swanson and F. Clearfield (eds.), Agricultural policy and the environment: Iron fist or open hand. Soil and Water Conservation Society, Ankeny, IA.

Napier, T., C. Thraen, and S. McClaskie. 1988. Adoption of soil conservation practices by farmers in erosion-prone areas of Ohio: The application of logit modeling. Soc. Nat. Resour. (1): 109–129.

Napier, T. and A. Napier. 1991. Perceptions of conservation compliance among farmers in a highly erodible area of Ohio. J. Soil Water Conserv. 46(3):220–224.

Napier, T. and D. Brown. 1993. Factors affecting attitudes toward ground water pollution among Ohio farmers. J. Soil Water Conserv. 48(5):432–438.

Napier, T., S. McCarter, and J. McCarter. 1995. Willingness of Ohio landowner operators to participate in a wetlands trading system. J. Soil Water Conserv. 50(6):648–656.

National Academy of Sciences, 1993. Soil and Water Quality: An Agenda for Agriculture. National Academy Press, Washington D.C.

Nelson, L. 1936. National policies and rural social organization. Rural Sociol. 1(1):73–89.

Nelson, L. 1937. Soil conservation and human welfare. Utah Farmer 57:3, 11, 21.

Norris, P. and S. Batie. 1987. Virginia farmers' soil conservation decisions: An application of tobit analysis. Southern J. Agric. Econ. 19(1):79–90.

Nowak, P. 1982. The selling of soil conservation: A test of the voluntary approach, phase one final report. Department of Sociology, Iowa State University, Ames, IA.

Nowak, P. 1983. Adoption and diffusion of soil and water conservation practices. Rural Sociol. 3(2):83–91.

Nowak, P. 1985. The leadership crisis in conservation districts. J. Soil Water Conserv. 40(5):420, 422–425.

Nowak, P. 1987. The adoption of agricultural conservation technologies: economic and diffusion explanations. Rural Sociol. 52(2):208–220.

Nowak, P. 1992. Why farmers adopt production technology. J. Soil Water Conserv. 47(1):14–16.

Nowak, P. and P. Korsching. 1985. Conservation and tillage: revolutionary or evolutionary? J. Soil Water Conserv. 40:199–201.

Nowak, P., M. Schnepf, and R. Barnes. 1991. When conservation reserve program contracts expire: A national survey of farm owners and operators who have enrolled land in the conservation reserve. Soil and Water Conservation Society, Ankeny, IA.

Pampel, F. and J.C. van Es. 1977. Environmental quality and issues of adoption research. Rural Sociol. 42(1):57–71.

Rogers, E. 1960. Social Change in Rural Society. Appleton-Century-Crofts, New York.

Rogers, E. 1995. Diffusion of Innovations. 4th ed. Free Press, New York.

Ryan, B. and N. Gross. 1943. The diffusion of hybrid seed corn in two Iowa communities. Rural Sociol. 8:15–24.

Sanderson, D. 1942. Rural Sociology and Rural Social Organization. John Wiley & Sons, New York.

Schaller, N. 1992. Review of farming in nature's image: An ecological approach to agriculture. J. Soil Water Conserv. 47(3): 254.

Taylor, C. 1930. Rural Sociology. Harper & Row, New York.

Taylor C., B. Allin, and O. Baker (Ed.), 1938. Public purposes in soil use. pp. 47–59 in U.S. Department of Agriculture, Soils and Men: Yearbook of Agriculture, 1938. U.S. Government Printing Office, Washington D.C.

Taylor, D. and W. Miller. 1978. The adoption process and environmental innovations: a case study of a governmental project. Rural Sociol. 43(4): 634–648.

Thomas, J., H. Ladewig, and W. McIntosh. 1990. The adoption of integrated pest management practices among Texas cotton growers. Rural Sociol. 55(3):395–410.

U.S. Department of Agriculture. 1938. Soils and Men: Yearbook of Agriculture, 1938. U.S. Government Printing Office, Washington D.C.

U.S. Department of Agriculture, Soil Conservation Service. 1989. Summary Report: 1987 National Resources Inventory. Statistical Bulletin 790. U.S. Department of Agriculture, Washington D.C.

U.S. General Accounting Office. 1977. To Protect Tomorrow's Food Supply, Soil Conservation Needs Priority Attention. Comptroller General of the United States, Washington D.C.

Valente, T. and E. Rogers. 1993. The rise and fall of rural sociological research on the diffusion of innovations: the Ryan and Gross paradigm. Paper presented at the 1993 Annual Meeting of the Midwest Sociological Society, Chicago, IL.

van Es, J. 1983. The adoption/diffusion tradition applied to resource conservation: inappropriate use of existing knowledge. Rural Sociol. 3(2):76–82.

Wagener, D. 1987. Sociological factors influencing the decision of Iowa farmers to adopt needed soil conservation practices. Unpublished Ph.D. dissertation. Department of Sociology, Iowa State University, Ames, IA.

Wakely, R. 1936. Social and economic effects of soil erosion. Rural Sociol. 1(4):509 510.

Wilkinson, K. 1969. Special agency program accomplishment and community action styles: The case of watershed development. Rural Sociol. 34(1): 29–42.

Williams, R. 1984. Introduction to Statistics for Geographers and Earth Scientists. Macmillan Publishers, Ltd., London.

Wolf, S. and P. Nowak. 1995. Development of information intensive agrichemical management services in Wisconsin. Environ. Manage. 19(3): 371–382.

Wrigley, N. 1979. Statistical Applications in the Spatial Sciences. Pion Limited, London.

Wynne, W. 1943. Culture of a contemporary rural community: Harmony, Georgia. Rural Life Studies: 6, Bureau of Agricultural Economics, U.S. Department of Agriculture, Washington, D.C.

Yates, F. 1981. Sampling Methods for Censuses and Surveys. Charles Griffin and Co., Ltd., High Wycombe, England.

The Connection Between Soil Conservation and Sustainable Agriculture

Dennis Keeney and Richard Cruse

Introduction

Soil erosion and its effects on crop production have been of concern to agricultural scientists since the turn of the century and particularly since the beginning of extensive land conservation during World War I. Wheat production for the war effort was a primary concern (Allin and Foster, 1940). A. R. Whitson and T. J. Dunnewald (cited in Johnson, 1991) wrote in 1916 that there were two major kinds of soil erosion damage: loss of fertility due to the selective removal of soil organic matter and the fine silt fraction, and development of gullies and ravines which "cut the fields so they cannot be cultivated."

In 1940, between the clouds of the Dust Bowl and WWII and before the widespread use of modern fertilizers, H. H. Bennett (1940) wrote: "Examined in the light of scientific knowledge, soil depletion is no simple process. It can result from the extraction of chemical elements from the soil, from the breakdown of soil structure, or from the actual removal of topsoil. Crops gradually remove the elements of fertility from the soil; methods of tillage and rotation have an important effect upon soil structure; and erosion by wind or water removes the entire body of the soil." In this statement, more than 55 years ago, Bennett captured much of the concept of soil quality, and recognized that soil erosion is one of the major soil depletion factors associated with agriculture. Dr. Bennett's view of agriculture is one that most would recognize today as a key part of sustainable agriculture. In his public address in 1955 on the eighteenth year of the Coon Creek (Wisconsin) Watershed establishment, he said "we could cure most or all of the difficulties which have plagued our production and marketing operations in recent years through the use of the land according to its capability" (Johnson, 1991).

Bennett then presented his interpretation of conservation practices, most if not all of which would be considered as sustainable (within the limits of the technologies available in 1955), viz: "Nature protects the land by clothing it with adaptable protective cover." He conceded that "we need some clean tillage for wheat, corn, tobacco and potatoes, but even here we should strive for as much cover as may be practicable through the use of such systems of modern soil conservation as seasonal cover cropping the stubble-mulching." But he also recognized that this concept of land use based on soil capability was not widely accepted. He noted the increasing amount of fields subjected to multiple tillage operations each year for seed bed preparation and weed control and the lack of protective vegetative cover for the land. Aldo Leopold also became involved in the Coon Creek watershed, particularly the revegetation and game management aspects. He emphasized the need for further integrating sustainable land management practices with soil conservation (Meine, 1988).

Defining Sustainable Agriculture

There have been many attempts to understand the sustainable agriculture concept. Definitions have ranged from practices to philosophies to visions (Keeney, 1990, 1995; Neher, 1992; Schaller, 1993). Neher (1992) indicates that no precise set formula applies to all situations, and that the future generations will be the best ones to judge if sustainability is being practiced currently. Schaller (1993) attempted to separate means and ends. Sustainable practices are the means to achieve the vision. Sometimes the means pull together and other times they are in conflict (e.g., the issue of conservation-tillage or no-tillage and dependence on herbicides as opposed to long-term rotations).

There are other agendas, at least in the United States, that involve sustainable agriculture. Sustainable agriculture has been successfully used as the base for a populist postindustrial movement to reverse the trends to industrial farming. This movement has influenced federal farm bill and research title provisions. Conversely, industrial supporters of sustainable agriculture have equated sustainability with the need for technology to meet future export demands and food needs (DowElanco, 1994; Avery, 1995).

Soil Conservation and Sustainable Agriculture

Seemingly, many modern agricultural technologies developed and promoted through public and private research, extension, policy development, and trade agreements have had major negative impacts on soil quality and soil erosion. The use of fertilizers to help overcome loss of soil productivity from erosion and degradation, the ever-increasing size of soil tillage and harvest equipment, and pesticides that minimize the need for crop rotation have caused today's agriculture to run counter to the soil protection ethic promoted by Bennett. Even more disturbing are indications that the many technological innovations of the past three decades have decreased stability of agricultural systems, making them more susceptible to energy, pest, and weather fluctuations (Robinson et al., 1994).

The "divorce" was never complete, however. Many leaders continued to promote soil conservation practices. A complete litany of local, state, and national leaders is impossible, and any attempt to develop a list would be incomplete. The torch continues to burn brightly. Modern day leaders of NRCS (National Resources Conservation Service) emphasize more than ever the need for land stewardship (Johnson, 1994).

Further, off-site costs of soil erosion need strong consideration. Heimlich (1991) estimated off-site costs of erosion at over $7 billion per year. Crosson (1991a) argues that the erosion-induced costs of soil productivity are not nearly the threat to U.S. and Canadian agriculture as are off-site environmental costs including sediment damage to water quality and habitat losses from drainage of wetlands. In contrast, Laflen et al. (1990) support the more commonly held view that soil erosion is a major threat to sustainable agriculture.

U.S. agriculture became more industrialized and grain production intensified in the early 1950s and soil loss from wind and water became an endemic problem nationwide (CAST, 1982; Crosson, 1991b). The separation of sustainability and soil conservation from agriculture occurred when soil came to be regarded as a commodity rather than an irreplaceable resource (e.g., Friend, 1992). The T (soil loss tolerance) concept was developed and promoted and soil loss standards set. Seldom has such an important policy been based on such a dearth of defendable data. Larson (1981) suggested a dual concept for T, of a functional and a political T value. This concept was never considered further, but might be worthy of debate if soil erosion and sustainable agriculture policies are to be brought together. While the meaning of agricultural sustainability and "sustainable practices" continues to be debated on a short-term, political (i.e., 2-4 year) basis, the strength of scientific opinion is that soil formation rates in many situations are so slow as to be negligible, and that many soils are nonrenewable within the terms of human lifespan (Friend, 1992).

The above discussion implies that row crop farming is not sustainable as defined by the American Society of Agronomy (1989). Their definition, "A sustainable agriculture is one that, over the long run, enhances environmental quality and the resource base on which agriculture depends . . ." would set soil loss at less than the rate of formation. And it implies an agriculture that incorporates soil building practices and increased soil quality.

Thus agriculture is faced with a classic trade-off, how to grow the erosive row crops with minimal damage to the environment and especially to the soils resource base. We believe that the principles of sustainability remain untested in most of the current soil conservation practices. The major exception is the Conservation Reserve Program, a program that may not be sustainable given the whims of politics and the pressures on the federal budget.

In the past 50 years, the primary goal of soil conservation efforts has been to keep soil from moving downslope or off-site while promoting increased productivity and efficiency of the agricultural system, or what Leopold (1949) termed improving the pump but not the well. This has been done by replacing landscape attributes such as permanent grassland, woodlands, and long-term rotation with continuous annual row crops and minimal return of crop residues and organic fertilizers to the soil. The latter protected soil from erosion and enhanced soil quality. The results have been: (1) a several-fold expansion in erosive row crops; (2) a marked increase in fertilizer and pesticide use; (3) frequent years of crop surpluses; (4) a federal policy to support prices with the result that crop production and use of inputs is expanded; and (5) a research establishment designed around enhancement of agricultural productivity and efficiency.

There have been a few "ticks" in the trend, largely based on federal government-supported programs to remove land from production. The conservation reserve program (CRP) is the largest and most successful to date. The CRP program has significantly reduced soil loss in sensitive areas and has shown that moving landscapes out of row crops can have large positive benefits such as improved water quality and wildlife habitat. The CRP has come at high cost to the public purse and has accelerated the depopulation of affected rural areas.

Several technologies and policies have recently joined to enhance the rapid increase in conservation tillage practices. Conservation compliance, initiated in the 1985 Food Security Act, rapidly moved clean-tilled row crop farming on erodible land to systems that left erosion controlling crop residues on the soil during most of the year. Acceptance has been enhanced by the improved efficiency of well-managed no-till and other conservation-till systems compared to clean-till systems. An important technology fostering no-till has been the availability of selective herbicides that permits weed control without secondary tillage. Time and energy are saved, and government price supports, by meeting compliance, is assured.

Economic studies have shown that conservation tillage systems also result in returns comparable to conventional tillage systems (Fox et al., 1991; Weersink et al., 1992). The enhanced efficiency has been translated into more land farmed per hour of labor, resulting in fewer farms and farmers. The use of conservation tillage is claimed by NRCS to have reduced erosion by more than 66% and conserved more than 1 billion tons of soil (USDA-SCS, 1994).

Soil Quality and Sustainable Agriculture

The Soil Science Society of America (1995) defines soil quality as "The capacity of a specific kind of soil to function, within the natural or managed ecosystem boundaries, to sustain plant and animal productivity, maintain or enhance water and air quality, and support human health and habitation." Soil quality is closely related to issues of soil conservation, and to sustainable agriculture (Karlen et al., 1990; NRC, 1993; Larson and Pierce, 1994; Warkentin, 1995). The 1993 report by the Board on Agriculture Committee on Long-Range Soil and Water Conservation (NRC, 1993) has brought new awareness to this concept and to the relation of the soil to the environment and to the long-term sustainability of the soil resource. It concerns issues central to environmental soil science including soil compaction, soil salinization, soil organic matter and associated biological factors, soil structure and attendant physical attributes.

The NRC report is important because it relates the impacts of soil quality on environment and production and also considers the relation of agricultural practices and policies to soil quality. Soil quality issues overlap many of the issues of sustainable agriculture. This is likely the most important interface of soil science to sustainable agriculture. As understanding is gained, the issues relating soil quality to sustainable agriculture will become much clearer.

The NRC (1993) report recommends (p. 201) that the soil quality concept should guide the recommendations for the use of conservation practices and for federal targeting of programs in resource conservation. It also emphasizes the importance of a holistic approach to the policy recommendations.

Soil Conservation Technologies, Policies, and Sustainable Agriculture

The development of policies and technologies that expand production of crops such as corn on highly erodible lands is a prime example of our economic system rewarding farmers for using unsustainable farm management. Abler and Shortle (1995) point out that these innovations can reduce marginal cost and increase output demand because of lower prices. If crop production increases sufficiently, the total use of potentially polluting inputs as well as soil erosion would actually increase. Thus, the current push for more efficient technologies needs to be viewed cautiously. Rapidly developing technologies that must be considered with regard to their impacts on soil conservation and on sustainable agriculture include precision farming and biotechnology. Government policies to be viewed with similar caution include the strong emphasis on exports.

Soil Science, the Landscape, and Sustainability

The concept of designing agricultural practices for the landscape is not particularly new, but has received renewed attention recently. The recognized failure of in-field approaches to solve environmental problems opened opportunities for a strong interface between sustainable agriculture and soil conservation when the landscape is considered. Landscape approaches include use of conservation easements for highly erodible land, vegetated buffer strips, strip intercropping, placement of livestock production areas to minimize water pollution and odor problems, and wetlands to remove contaminants from runoff. The combination of good science and a land use policy dedicated to environmental improvements and economical crop production offers widespread benefits. These include water quality enhancement, improved soil quality, ecotourism, and higher returns on land best suited for crop production. While much is promised from the landscape planning concept, delivery on the promises will require diligent effort. There are many hindrances, first and foremost the political and property rights issues which make landscape planning difficult at best. And, as well-stated by Stanley (1994), application of science or technology to solve ecosystem level resource or conservation problems often is doomed to failure because of the anthropogenic objectives of management. However, Peterson et al. (1993) have demonstrated the power of an agroecosystem approach to soil and crop management research at the landscape level without losing the ability to detect cause and effect, while providing technology transfer from the researcher to the user.

The Future for Soil Conservation Research Related to Sustainability

As agriculture moves into the postindustrial age (some say this is approaching sustainable agriculture) the link between soil conservation and sustainability will be even more critical. Soil conservationists will find it difficult to isolate the reductionist and on-field approaches of the past. Can the vision of Hugh Bennett be reincarnated? What role will the new soil conservation research agenda play in sustainability?

Four main issues should be considered. These are the application of technology, defining and building soil quality, holistic landscape approaches, and implementation of new farming systems.

Application of Technology

The information technology and biotechnology advances that began several years ago and are now in full swing offer many opportunities for advancing a soil-conserving sustainable agriculture. However, the inevitable outcome of many technological advances that increase efficiency is more concentration of agriculture in the industrial mode. This is not the fault of a technology per se. Rather, the technology (or packet of technologies) is being marginally applied for the benefit of farmers but with much larger benefit for investors and other stakeholders.

Precision farming. Precision farming (also called site-specific farming) permits management of production factors on a very small land area, allowing this area to be managed as a separate unit (Christensen and Krause, 1995). The technologies to support precision farming are developing rapidly. Better management of inputs would provide benefits to the landscape, and likely would lessen resource damage. However, it is not yet clear if the reduced inputs and/or higher outputs will be sufficient to cause its widespread adoption. With current marketing strategies, precision farming would more closely fit to the industrial than the sustainable agricultural model. If so, it would also meet the criteria of Abler and Shortle (1995) that might actually increase the total output of pollutants.

Plant breeding. Plant breeding is another technology that offers much promise. New crops bred to provide season-long soil cover may be able to mesh soil conservation and sustainable agriculture. Improved fiber crops and perennials are also possible. However, current industrial application of plant breeding, including biotechnology, appears to be concentrating on improved food and feed varieties of cultivated crops, particularly those that provide a strong profit to the industry (Duvick, 1995).

Defining and Building Soil Quality

Soil quality likely offers the most promise for close linkage of soil conservation principles and sustainable agriculture principles. Therefore, exploratory and adaptive research in soil quality should be encouraged. Importantly, the landscape basis of soil quality enhancement should be evaluated and the socioeconomic and farmer assessment of soil quality must be placed in context with the agronomic aspects (Romig et al., 1995). Warkentin (1995) puts it well: "The exciting work of using soil quality concepts for sustainable production and environmental protection has just begun."

Holistic Landscape Approaches

The soil conservation professional and researcher must involve people, particularly those who own and manage the land, if sustainable agriculture is to be achieved. It is clear that they need to emphasize conserving practices on the landscape-scale. Research needs are cross-disciplinary. In most cases, the sustainable landscape technologies (e.g., vegetated buffer strips) will not be cost-effective (with a few possible exceptions), so incentives and perhaps regulations must be required for their adoption. Team approaches will be necessary and will need support from a wide array of technical and social sciences as well as end-users and policy-makers.

Farming Systems

A key to soil conservation improvements is development of new farming systems which go beyond efficiency or substitution approaches and emphasize redesigning the way things are done. This will be long-term, high-risk research and development, something that is not easily supported by conventional funding sources, land grant universities, or federal Agricultural Research Service (ARS) programs.

Summary

The approaches used to conserve soil relative to those needed to assure a sustainable agriculture have divergent goals. The main objective of past soil conservation technologies has been to minimize soil loss while maximizing efficiency of row crop production. This has not necessarily promoted sustainable agriculture, and in many situation has increased total agricultural chemical use and soil loss by making more land available for row crop agriculture. Soil quality, a concept still in its infancy in definition, offers a major possibility for linking soil conservation and sustainable agriculture.

References

Abler, D.G., and J.S. Shortle. 1995. Technology as an agricultural pollution control policy. Am. J. Agric. Econ. 77:20–32.

Allin, B.W., and E.A. Foster. 1940. The challenge of conservation. pp. 416–428. In: Farmers in a Changing World. United States Department of Agriculture.

American Society of Agronomy. 1989. Decision reached on sustainable agriculture. Agron. News, January. p. 15.

American Society of Agronomy. 1995. SSSA statement on soil quality. Agron. News. p. 7.

Avery, D.T. 1995. Saving the planet with pesticides and plastic: The environmental triumph of high–yield farming. Hudson Institute. Indianapolis, IN.

Bennett, H.H. 1940. Our soil can be saved. pp. 429–440. In: Farmers in a Changing World. United States Department of Agriculture.

Christensen, L. and K. Krause. 1995. Precision farming: Harnessing technology. Agric. Outlook. May, 1995. pp. 18–19.

Council for Agricultural Science and Technology (CAST). 1982. Soil erosion: Its agricultural, environmental and socio–economic implications. CAST Report 92. Jan. 1982.

Crosson, P. 1991a. Sustainable agriculture in North America: Issues and challenges. Can. J. Agric. Econ. 39:553–565.

Crosson, P.R. 1991b. Cropland and soils: Past performance and policy challenges. pp. 169–203. In: America's renewable resources: Historical trends and current challenges. Resources for the Future. Washington, D.C.

DowElanco. 1994. Sustainable agriculture 1994. The Bottom Line. DowElanco. Indianapolis, IN.

Duvick, D.N. 1995. Biotechnology is compatable with sustainable agriculture. J. Agric. Environ. Ethics 8:112–125.

Fox, G., A. Weersink, G. Sarwar, S. Duff, and B. Deen. 1991. Comparative economics of alternative agricultural production systems: A review. Northeastern J. Agric. Res. Econ. 20:124–142.

Friend, J.A. 1992. Achieving soil sustainability. J. Soil Water Conserv. 47:156–157.

Heimlich, R.E. 1991. Soil erosion and conservation policies in the United States. pp. 59–89. In: N. Henley, (ed.) Farming and the countryside: An economic analysis of external costs and benefits. C.A.B. International, UK.

Johnson, L.C. 1991. Soil conservation in Wisconsin: Birth to rebirth. The Department of Soil Science, University of Wisconsin, Madison.

Johnson, P. 1994. Taking stock. In: Gaining ground: Soil conservation achievements of U.S. farmers. United States Department of Agriculture. Soil Conservation Service. August 1994.

Karlen, D.L., D.C. Erbach, T.C. Kaspar, T.S. Colvin, E.C. Berry, and D.R. Timmons. 1990. Soil tilth: A review of past perceptions and future needs. Soil Sci. Soc. Am. J. 54:153–161.

Keeney, D.R. 1990. Sustainable agriculture: Definitions and concepts. J. Prod. Agric. 3:281–285.

Keeney, D.R. 1995. Sustainable agriculture: A vision for Iowa. Leopold Letter 7(2):3,11.

Laflen, J.M., R. Lal, and S.A. El–Swaify. 1990. Soil erosion and a sustainable agriculture. pp. 569–581. In: C. W. Edwards et al., (eds.) Sustainable agricultural systems. Soil and Water Conservation Society. Ankeny, IA.

Larson, W.E. 1981. Protecting the soil resource base. J. Soil Water Conserv. 36:13–16.

Larson, W.E. and F.J. Pierce. 1994. The dynamics of soil quality as a measure of sustainable management. In: J.W. Doran, D.C. Coleman, D.F. Bezdicek, and B.A. Stewart (eds.) Defining soil quality for a sustainable environment. Soil Science Society of America Special Publication 35:37–51. ASA, Madison, WI.

Leopold, A. 1949. A Sand County Almanac and Sketches Here and There. Oxford University Press, Oxford, U.K.

Meine, C. 1988. Aldo Leopold: His Life and Work. University of Wisconsin Press, Madison.

National Research Council. 1993. Soil and water quality: An agenda for agriculture. Committee on Long-Range Soil and Water Conservation. Board on Agriculture. National Academy Press, Washington, D.C.

Neher, D. 1992. Ecological sustainability in agricultural systems: Definition and measurement. pp. 51–61. In: Integrating sustainable agriculture, ecology, and environmental policy. Richard K. Olson, (ed.) Food Products Press, Binghamton, NY.

Peterson, G.A., D.G. Westfall, and C.V. Cole. 1993. Agroecosystem approach to soil and crop management research. Soil Sci. Soc. Am. J. 57:1354–1360.

Robinson, C.A., R.M. Cruse, and K.A. Kohler. 1994. Soil management. pp. 109–134. In: Sustainable Agriculture Systems. J. Hatfield and D.L. Karlen (eds.) Lewis Publishers, Boca Raton, FL.

Romig, D.E., M.J. Garland, R.F. Harris, and K. McSweeney. 1995. How farmers assess soil health and quality. J. Soil Water Conserv. 50:229–236.

Schaller, N. 1993. The concept of agricultural sustainability. Agric. Ecosystems Environ. 46:89–97.

Stanley, T.R., Jr. 1994. Ecosystem management and the arrogance of humanism. Conserv. Biol. 9:255–262.

USDA-SCS. 1994. Gaining ground: Soil conservation achievements of U. S. farmers. USDA-SCS. Washington D.C.

Warkentin, B.P. 1995. The changing concept of soil quality. J. Soil Water Conserv. 50: 26–228.

Weersink, A., M. Walker, C. Swanton, and J. Shaw. 1992. Economic comparison of alternative tillage systems under risk. Can. J. Agric. Econ. 40:199–217.

Whitson, A.R. and T.J. Dunnewald. 1916. Keeping our hillsides from washing. Wisconsin Agricultural Experiment Station Bulletin No. 272. Cited in Leonard C. Johnson. 1991. Soil conservation in Wisconsin: Birth to rebirth. The Department of Soil Science, University of Wisconsin, Madison.

Policy and Government Programs in Soil and Water Conservation

W.E. Larson, M.J. Mausbach, B.L. Schmidt, and P. Crosson

Introduction

Research and educational programs have had a major influence on the formulation and enactment of government policies in soil and water conservation. Likewise government policies at both the federal and state level have influenced the kind and emphasis of research and education programs. In short, policies and research programs have been interactive, which has strengthened both activities. In some cases, a policy or program was proposed or enacted but was not successfully carried out until the needed research had been done. In other cases, research showed the need for a government program which was later enacted.

The Land Grant College Act in 1862 was the first major legislation by the Congress that influenced soil and water research and education programs. This established the Land Grant Universities and set the tone for future legislation. Many legislative acts since that time have strengthened the cooperative efforts between legislators and researchers and educators.

A constant dialogue between policy-makers and researchers is needed. Effective communication can ensure that current information is used in the development of policies and can guide the selection of the most pressing research needs. Effective communication among legislators, researchers, educators, and public interest groups has helped these groups to become closer in the last decade or so.

It is the purpose of this manuscript to review the interaction between public policy and research and to comment on the contributions from both groups.

Early Legislation

Colonial and early U.S. leaders were concerned about soil erosion and land degradation in the states east of the Appalachian mountains (Gray, 1933; Betts, 1953). They clearly saw the need to conserve the soils of the 13 original states if the agrarian economy of that time was to be preserved. However, it wasn't until May 15, 1862 that President Abraham Lincoln signed into law an act of Congress establishing the United States Department of Agriculture (USDA), with a Commissioner appointed by the President (Table 1). On Feb-

Table 1. Major Legislation Relative to Soil and Water Resources, 1862–1953

Date	Act	Purpose
1862	U.S. Dept. of Agr. established	Recognition of importance of agriculture to nation's welfare
1862	Land Grant College	Public lands given to states to encourage establishment of agriculture and mechanic arts
1887	Hatch Act	Established state agricultural experiment stations
1889	Sec. of Agr. made member of President's Cabinet	Recognition of the importance of agriculture to nation's welfare
1890	Extension of Land Grant College	Extended Act to historically black colleges
1914	Smith–Lever Act	Established state Extension Service
1935	Soil Conservation Service	Established an agency in USDA to foster soil and water conservation
1953	Agr. Res. Service	Transferred all federal soil and research to ARS, and all soil survey to SCS

ruary 9, 1889, the chief executive of the USDA was made Secretary of Agriculture with full cabinet status.

Passage in 1862 of the Land Grant College Act (Morrill Act) was the first major act by the U.S. Congress that significantly affected soil and water research. The Morrill Act encouraged states to establish colleges of agriculture and the mechanic arts and was extended in 1890 to include historically black colleges. Under the law, states received grants consisting of vast amounts of public lands to be managed and sold in support of the colleges. The Hatch Act of 1887 established the agricultural experiment stations in association with the land-grant universities. Nearly 30 years passed before passage of the Smith-Lever Act of 1914, which established the Extension Service to enhance transfer of the findings of research to farm producers. Research in soil and water has been concurrently conducted by a number of agencies in the federal government dating back to the establishment of the USDA, and by the State Agricultural Experiment Station (SAES).

In 1928, Dr. H.H. Bennett, a soil scientist from North Carolina, published a bulletin (along with R. Chapline) entitled 'Soil Erosion, a National Menace' (Bennett and Chapline, 1928) and began a campaign to obtain federal support for the study of erosion and means to control it. The drought and Dust Bowl of the early 1930s encouraged Congress to support Dr. Bennett's crusade and in 1933 a Soil Erosion Service was established within the Department of the Interior. In 1935 the Soil Conservation Service (SCS) was established as a permanent agency of the Department of Agriculture, with Dr. Bennett as its Chief. The SCS had a research division and 10 Erosion Experiment Stations (Middleton et al., 1932). A reorganization of the USDA in 1994 renamed SCS, the Natural Resources Conservation Service (NRCS).

Since 1953, soil and water research for cropland has been concentrated in the Agricultural Research Service (ARS), with the exception of economic research, which is in the Economic Research Service (ERS) of the USDA. The Cooperative State, Research, Education, and Extension Service (CSREES) coordinates research within the SAES and administers federal research grant programs. Research for forest lands has been the responsibility of the Forest Service (FS).

Soil Surveys

Soil surveys provide an inventory of the quality and aerial extent of soils and landscapes. As such, they form the cornerstone for land use and management decisions. Soil surveys have been used in implementing nearly all government programs in soil and water. The authorization for making soil surveys by the federal government dates to the establishment of the USDA. Language in the Act establishing the USDA assigned the USDA responsibility "to acquire and to diffuse among the people of the United States useful information on subjects connected with agriculture, rural development, aquaculture, and human nutrition, in most general and comprehensive sense of those terms . . ."*

The Agriculture Appropriation Act of 1896 provided more specific authority and authorized the "investigation of the relation of soils to climate and organic life; for the investigation of the texture and composition of soils, in the field and laboratory . . ." by the Division of Agricultural Soils. However, the origin of soil survey was set in 1899 when the first appropriations were received and the first report of field operations of the Division of Soils was published (Simonson, 1987). With the birth of the SCS in 1935, the SCS started a second soil survey to support the conservation efforts of the new agency. The SCS surveys concentrated on mapping individual farms in support of requests for conservation planning while the Bureau of Plant Industry, Soils, and Agricultural Engineering (BPISAE) surveys were designed for county-wide mapping. From 1935 to 1951 the two surveys were both operational. The two surveys were combined in 1952 and the SCS was given national leadership within the USDA for soil survey. From this combination, the National Cooperative Soil Survey (NCSS) program was developed and the "modern" soil

*U.S.A. 2001 Establishment of Department, dated May 15, 1882.

survey program began. The NCSS is a partnership among federal, state, and local agencies. Usually leadership for coordination within a state was assumed by the Agricultural Experiment Stations, where much of the research was performed.

During the period 1935 to 1952, the Land Capability Classification (LCC) system was developed and has been used extensively for interpreting soil survey information for conservation planning purposes. After the soil surveys were combined, the criteria and classification of the LCC were clarified (Klingebeil, 1991). The SCS and NCSS are authorized to provide technical assistance in the use of soil survey information to include interpretations for rural and urban uses of soils (Klingebeil, 1991; Soil Survey Staff, 1992).

Soil surveys are used extensively in federal, state and local policies and regulations. In 1966, the Soil Science Society of America published the book "Soil Surveys and Land Use Planning" (Bartelli et al., 1966). This book details uses of soil survey information ranging from zoning ordinances, building of highways, septic tank absorption fields, land use (urban) planning, and equalization of tax assessments. At the federal level, soil surveys are part of the Prime Farmland Protection Policy Act (included in the 1981 Farm Bill), Surface Mining Control Act, the Highly Erodible Land (HEL) and Swampbuster title of the Food Security Act (FSA) of 1985, the Wetland Reserve and Conservation Reserve Programs of the FSA and Food, Agriculture, and Trade Act of 1990 (FACTA), and the Resource Conservation Act (RCA).

In 1955, Fairfax County, Virginia, was the first county to hire a soil scientist for urban interpretations of soil survey information (Obenshain, 1966; Larry Johnson, soil scientist, Fairfax County, personal communication, 1995). Fairfax County continues to employ two full-time soil scientists, a geologist, and a cartographer. They interpret soil survey information for septic tank absorption fields, high water tables, wet soil problems, soil slippage or instability, and other factors related to development of an area. The staff also investigates hazardous waste spills in the county. They are involved in wetland issues with respect to permitting and mitigation. They developed a radon interpretations map of the county. Another example of use of soil survey information in local communities is the Tahoe Regional Planning Agency, which is regulating and prohibiting residential and urban development on sensitive lands within the Lake Tahoe region via a soil-based individual Parcel Evaluation System. This system rates parcels on the potential for erosion and runoff hazards and the potential for reestablishment of vegetation (Shellhorn, 1993). Many states use soil survey information to develop equitable procedures for taxing land values. Soil surveys provide uniform data across a county, the usual taxation unit, but also provide for equalization among counties in a state. Often, state agricultural experiment stations have estimated the productive capacities of soils across the state (Anderson et al., 1992).

The Surface Mining Control and Reclamation Act of 1977* set forth special

---

*Public Law 95-87, 30 U.S.A. 1201 et seq.

environmental protection performance, reclamation, and design standards for surface coal mining and reclamation operations on prime farmland. The definition of prime farmland is based entirely on soil properties collected in the NCSS and stored in the Soil Interpretations Record (SIR) database. The surface mine law requires states to adopt reclamation procedures such that the productivity of the prime farmland will be restored to the productivity of the land before mining activities.

The Farm Land Protection Policy Act of 1985 requires federal agencies to identify and take into account the adverse effects of their programs on the preservation of prime or unique farmlands, to consider alternative actions, as appropriate, that could lessen adverse effects, and to ensure that their programs, to the extent practical, are compatible with state and units of local government and private programs and policies to protect prime and unique farmland. The amendments in the FSA require the USDA to make an annual report to Congress on these effects of federal programs on prime and unique farmlands.

The RCA of 1977 directs the Secretary of Agriculture to conduct a continuing appraisal of the soil, water, and related resources on nonfederal land of the U.S. The Act further stipulates that programs administered by the Secretary of Agriculture for the conservation of soil, water, and related resources be responsive to the long-term needs of the country as determined by the appraisal. The USDA issued the first appraisal in 1981, the second in 1989, and is presently working on a joint appraisal and strategic plan due in 1997. The FSA of 1985 required RCA appraisals through the year 2005 (Soil Conservation Service Staff, 1989). These appraisals draw heavily on the soil survey information in the National Resources Inventory (NRI) (Soil Conservation Staff, 1994), conducted every five years. Models (e.g., USLE, WEQ, EPIC) employed in the appraisal use information in the SIR.

With the increasing need and use of soil survey information, a revised approach to soil inventories is needed. Additional landscape information, the scale of mapping for local needs, and means to aggregate and disaggregate data as the need arises, all need further attention. Increased cooperation between the soil survey activities by the NRCS and the states is desirable to ensure that local needs are served. The need to make soil surveys and related information more easily used and the tremendous opportunities for new uses of soil survey has dramatically increased interest in digitizing the information for computer recall and interpretation. Both federal and state support for soil surveys has waned in the last decade or so. While some states have received state-appropriated monies, the total level of funding is inadequate.

In summary, soil surveys have provided a quantitative inventory of the characteristics, extent, and location of the nations' soils. Soil surveys have been used in a host of ways for guiding use and management of the nations' lands. Information based on soil surveys has been increasingly and effectively used in implementing and enforcing government programs, and in assessing the ability of the nation to sustain an ample food supply and protect the environment.

Erosion Prediction

Perhaps no other research development has had more influence on erosion control technology than development of procedures for erosion prediction. Erosion prediction was initiated for guiding conservation planning. The beginnings of what eventually was called the Universal Soil Loss Equation (USLE) started in the 1930s and early 1940s from data collected at the Erosion Experiment Stations. Early leaders included Zingg (1940), Smith and Whitt (1947), Browning et al. (1947), and Musgrave (1947). In 1954, the ARS formed a laboratory at Purdue University to analyze all the available data and develop prediction procedures. This work headed by Wischmeier and Smith (1965) developed what is now called the USLE. It was landmark research. While the USLE is still widely used, the need for improved prediction has led to the development of the Revised Universal Soil Loss Equation (RUSLE) (Renard et al., 1991) and Water Erosion Prediction Program (WEPP) (Laflen, 1997.)

The USLE, along with soil loss tolerance or T-values, has been used as the standard for judging the suitability of a set of conservation practices to achieve acceptable erosion amounts. T-values, were defined by Wischmeier and Smith (1978) as "the maximum level of soil erosion that will permit a high level of crop productivity to be sustained economically and indefinitely." In practice, a soil conservationist designs a system for a given field, and then uses the USLE to check if the estimated erosion amounts exceed the estimated T-values. If the estimated erosion exceeds T-value, additional conservation practices are added until the amounts are equal to or less than T. The USLE has been used in the RCA assessments to estimate the amounts of nationwide erosion and to determine how much land was eroding at unacceptable amounts. Factors in the USLE have been used to define and map highly erodible land (HEL) subject to conservation compliance provisions of the FSA of 1985. Likewise, the USLE is used to develop conservation systems on highly erodible land as required in the 1985, 1990, and 1996 Farm Bills. Thus, the USLE has been the cornerstone on which government soil conservation programs for the humid regions have been built.

Because the USLE was successfully used for estimating water erosion, a similar equation was needed for estimating wind erosion. Such an equation was developed (WEQ) (Skidmore and Woodruff, 1968) and has been used for conservation planning, for RCA assessments, and for erosion compliance. However, it is generally conceded that major improvements in the estimating procedures are needed. Currently a new wind erosion estimating procedure is being developed and promises to be a big improvement (Fryrear, 1998). It is known as the Wind Erosion Prediction System (WEPS).

No special congressional funding or directives have been issued over the years concerning erosion prediction technology, although it has remained a high research priority by the NRCS, ARS, and SAES. The 1996 Farm Bill requires publication of the USLE and any future modifications in the *Federal Register*.

Erosion Consequences

Since recorded history began, it has been obvious that significant soil erosion decreased the potential productivity of most soils (Lowdermilk, 1953). Reductions in soil productivity were related to removal of plant nutrients, reduced water-holding capacities, reduced rooting volumes, and many other causes. While the decreases in plant growth were obvious, the specific cause or causes were not clearly understood, and were not quantified. With passage of the 1977 RCA and preparation of the first assessment, policy leaders were asking researchers and analysts to quantify losses in potential productivity due to erosion. Researchers responded by using several different approaches (Pierce et al., 1984; Crosson, 1986; Crosson and Rosenberg, 1989). Pierce et al. (1984) showed that potential soil productivity losses from erosion at the amount estimated for 1977, if continued for the next 100 years would range from 1.7 to 7.8% for 15 Major Land Resource Areas in the Corn Belt. The losses by slope class, however, varied greatly, ranging up to 100% for slopes of 12 to 20%. Pierce et al. (1984) did not include losses from nutrient depletion and from ephemeral and gully erosion. The large losses on the steeper slopes reinforced the concept that special attention is needed for these highly erodible soils.

Recent research has shifted attention from on-site productivity losses due to erosion to off-site damages in most regions of the U.S. A number of researchers have analyzed the economic losses from both on-site and off-site damages (Clark et al., 1985; Crosson, 1986; Colacicco et al., 1989) and concluded that off-site damages were at least twice that of on-site damages. Colacicco et al. (1989) reported that on-site damages were greater than off-site damages in the Mountain and Southern Plains Regions (Figure 1). Off-site damages were greater in all other farm production regions. Colacicco et al. (1989) point out, however, that while overall productivity losses due to soil erosion may be small in some areas, they cannot be ignored. Research showing that off-site damages exceeds that from on-site damages has resulted in greater emphasis on practices that trap sediment at the edge of agricultural fields or along riparian areas. Vegetative filter strips and constructed earthen sediment traps are examples.

Because the estimated cost of off-site sediment damage is several times larger than on-site losses in productivity, the question can logically be raised as to the adequacy of current T-values. The original basis for assigning T-values is obscure (Johnson, 1987; Shertz, 1983) but was concerned with on-site

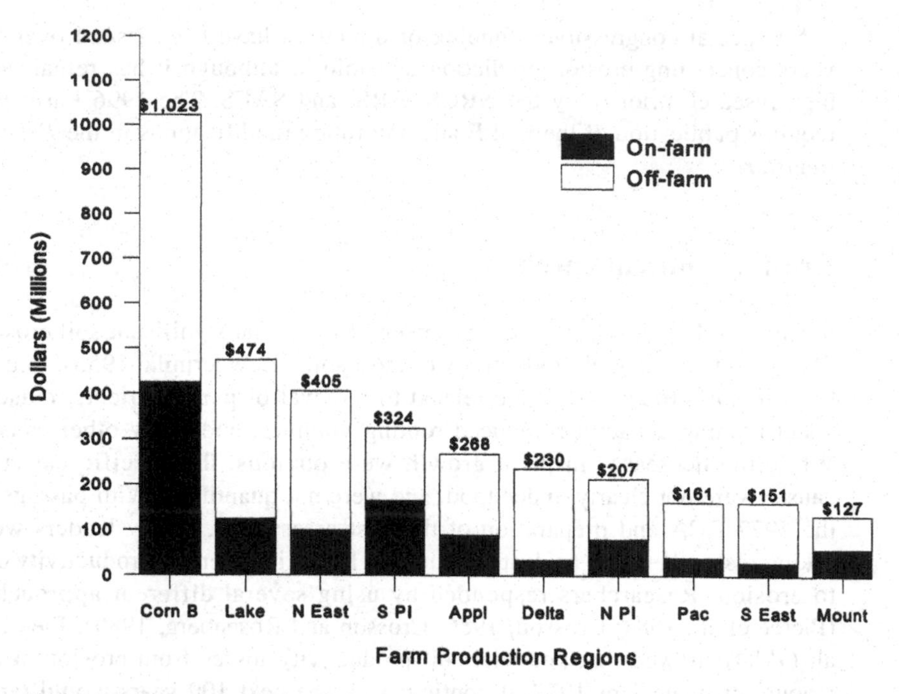

Figure 1. On–farm and off–farm damages from sheet, rill, and wind erosion. (From Colacicco et al., 1989).

productivity losses. The concept of vulnerability to productivity losses for different soils developed by Pierce et al. (1984) clearly points out that some soils can withstand considerably larger soil removal than the maximum value for T of 11.2 tons ha^{-1} $year^{-1}$ without appreciable losses in long-term nonreplaceable productivity. Both the USLE and WEQ predict erosion losses from a point, but do not necessarily have a direct relationship to off-site sediment amounts. Clearly a reassessment of how to quantify when a management unit is performing in an acceptable manner with respect to erosion control is needed. Currently in the conservation compliance section of the FACTA, a field is judged in compliance if an approved conservation plan is implemented. Alternative ways to judge whether erosion is at acceptable levels is discussed under the section on Cropland Acreage Reductions.

Because erosion is the major degrading force to the nation's lands, it is imperative that means of estimating the amount of erosion and its damaging consequences are available. The USLE and WEQ, and recent improvements, have provided means of estimation that have been used for guiding the development of government programs, for assessing damages from erosion, and for designating parcels of lands needing improved erosion control. A current research need is to develop management systems on an ecosystem or watershed basis that provides economic production and protects soil, water, and air quality. Improved site-specific identification of management units for delineating best land use and management is needed for both government and private uses.

Conservation Tillage

Improved tillage practices for soil and water conservation has been a major goal of research since the 1930s when the SCS (now NRCS) was founded. For erosion control, research quickly demonstrated the benefits of reducing the number of tillage operations, leaving the soil surface rough, and leaving all or part of the residues from the previous crop on the soil surface. While some early success was achieved for the Great Plains, material progress in getting conservation tillage on the land for the humid regions required the development of modern herbicides, improved tillage machinery, and improved nutrient management. In the last two decades progress has been rapid (Blevins et al., 1998; Allmaras et al., 1998).

Duley and Russell (1939; 1942) developed the stubble mulch tillage system for the Great Plains in the 1940s and 1950s. This basic system has since been called mulch tillage, ecofallow, and conservation tillage. Basically, the Duley and Russell system advocated using large sweeps to undercut the soil while retaining the crop residues on the surface. The system abandoned use of the moldboard plow that had long been the basic primary tillage tool. It wasn't until the 1950s and 1960s that significant progress was made in the development of forms of conservation tillage for the humid regions. Early systems included wheel-track planting, mulch-tillage, ridge-planting, and strip tillage (Larson et al., 1956). These systems relied on fewer tillage operations, rough soil surfaces, and/or residue cover.

In the last decade or so, various forms of conservation tillage have been put on the land. For row crops in the humid areas, the most popular forms have been no-till, ridge-till, and chisel-till. For example, in Iowa in 1991, 5.2% of the cropland in corn employed no-till, 1.5% used ridge-till, and 19.7% used mulch till. Similar percentages occurred in Minnesota (Allmaras et al., 1994). In 1991, over 70% of the cropland in the Cornbelt and Great Plains was tilled without use of the moldboard plow.

Recent federal programs have done much to encourage greater use of conservation tillage. The concern about erosion that was engendered by fence-row to fence-row production in the early 1970s, the passage of the RCA Act in 1977, and subsequent NRI surveys and reports therefrom, focused the public's attention on the need for better erosion control. For example, several states developed educational programs with the slogan "T by 2000" (Cooper and Bollman, 1989), with the goal to get all land in the state to reduce erosion to the T level by the year 2000. A current incentive is the conservation compliance provision of the 1990 Farm Bill. Without the conservation tillage developments of the last two decades, meeting the conservation compliance provisions of the 1990 Farm Bill would have been impossible on much of the cultivated cropland, because no other practical alternative is available that doesn't require changes in the crops produced or in land use. The conservation tillage developments were led by a wide variety of researchers and helped by many innovative crop producers and machinery companies (Blevins et al., 1998).

Tillage of the soil for crop production can influence the degradation of the soil by influencing organic matter levels, infiltration, erosion, crop residue management, compaction, and other characteristics. Conservation tillage has been a major factor in recent decreases in erosion amounts and in management of soils for sustainability. It has been a key management tool in the compliance provisions of the FACTA and FAIRA. The dual approach of the 1980s and 1990s of effective research and education programs, coupled with sound government information and incentive programs is, perhaps, one of the best examples of how to get improved management practices on the land.

Soil Quality

Soil quality and the ability of the soil to function with respect to environmental and land use needs is an important issue. The 1993 National Research Council Report on "Soil and Water Quality (National Research Council, 1993) states that "soil and water quality are inherently linked." The report concludes that soil quality must be considered as having equally high priority with water quality for future research needs. The Soil Science Society of America recently adopted an official definition of soil quality and held a symposium on soil quality research at its annual meetings in 1994 (Doran et al., 1994). The ARS and CSREES have funded research on soil quality.

The National Research Council (1993) suggests that "national policy should seek to (i) conserve and enhance soil quality as a fundamental first step to environmental improvement; (ii) increase nutrient, pesticide, and irrigation use efficiencies in farming systems; (iii) increase the resistance of farming systems to erosion and runoff; and (iv) make better use of field and landscape buffer zones." They go on to mention that national policies have focused too narrowly on erosion and that soil compaction, salinization, acidification, and loss of biological diversity are important soil quality concerns. Soil survey information is an essential link in evaluating soil quality over time and land use, since each soil has an inherent ability to function within a specific land use. Information on the static soil properties in the SIR are important for establishing baseline levels from which to monitor trends in soil quality in space, time, and land use. Our next challenge is to determine the most important soil properties that vary in time, space, and land use that can be used for monitoring changes in the quality of soil and related resources; and that can be used to estimate the effects of changes in soil quality on crop productivity, water quality, and other environmental concerns.

The concept of Precision Agriculture (doing the right thing, at the right time, in the right place in the right way) is appealing. It offers the promise of maximizing crop yields and profits, while protecting soil, water, air, and food resources. As world food reserves, particularly grains, continue to shrink, maximizing crop production while protecting the resources will be critical. Precision Agriculture is a rapidly expanding technology and is based on the concept that every taxonomic or other unit within a farm is different and needs to be managed differently for agronomic, economic, and environment reasons (Pierce and Sadler, 1997). Precision Agriculture has been spurred by major research and development advances in new computer, satellite, geographic information systems, and geopositioning systems technologies which permit increasingly accurate measurements of site locations on the landscape, and on-the-go adjustments of tillage, planting, chemical applications, and harvesting equipment. Research is needed to evaluate the effects of spatial variability of soils and landscape characteristics on soil and water quality and crop productivity, and the application of this knowledge to develop soil-specific management practices for use in profitable precision-farming systems. A need exists for improved and additional digitized soil and landscape information for use in machines that apply the technology. Site-specific information is needed for many other uses; including tax assessment, waste utilization, and environment, as well as for monitoring the state of our resources as is done in the RCA.

Currently, national leaders have recognized the need to quantify the quality and changes in the quality of soil (National Research Council, 1993). Perhaps, meaningful state and national policies on soil quality must await research developments.

Crop Production

Soil and crop productivity research over the past century has not only increased yields, but has also influenced soil and water conservation. Indirectly, crop productivity research has had a great influence on government soil and water policy, such as the increased production that has made necessary the set-aside programs that were first started in the 1930s.

The virgin soils of the U.S. were high in organic matter and nutrients. However, by the 1930s many of the soils of the U.S. had been severely depleted of both organic matter and nutrients (Haas and Evans, 1957). Prior to World War II (WWII), fertilizers were applied at low rates or not at all to most of the major crops in the U.S. Along with low nutrient use and low plant populations, the resultant crop and residue soil cover in the 1930s was relatively low, both during the crop season and during the noncrop period. These factors, along with droughts and poor tillage practices, contributed to the Dust Bowl in the Great Plains and excessive water erosion in the eastern and central U.S.

Technology and production facilities developed during WWII, particularly for nitrogen synthesis, became available to the fertilizer industry after the war, which resulted in ample supplies of fertilizer at low prices. As a result, fertilizer use increased dramatically until about 1976 and then started to level off. Since 1985, fertilizer use has been stable (Figure 2). Increased fertilizer use, new cultivars, and other improved management techniques resulted in plant population increases and yields of many of the row crops during the past one-half century, particularly corn. These increases resulted in much greater plant residue production and increased canopy protection of the soil surface during the growing season. The average U.S. corn yield in 1940 was 1.9 Mg ha-1, and in 1990 was 7.3 Mg ha-1. Based on data of Allmaras et al., (1998), the amount of aboveground biomass excluding grain was 3510 kg ha-1 in 1940 and was 7310 kg ha-1 in 1990. This would result in 65 and 87% cover at harvest time. If it is assumed that 50% of the crop residue was buried or decomposed before or during tillage the following spring, (i.e., 3510 and 7310 kg ha-1, respectively), then the residue cover following corn would be about 43 and 67% at planting time for the 1940 and 1990 examples, respectively. Without the present production of residues, modern conservation tillage would be much less effective. In the 1930s, corn populations were about 27,000 plants ha-1, whereas they are about 2 to 3 times that today. The greater populations of today, along with narrower row-widths, result in greater canopy cover during the growing season. Increased plant residues, if managed well, can also increase soil organic matter (Larson et al., 1972). Increased cover and soil organic matter has been a deterrent to both wind and water erosion.

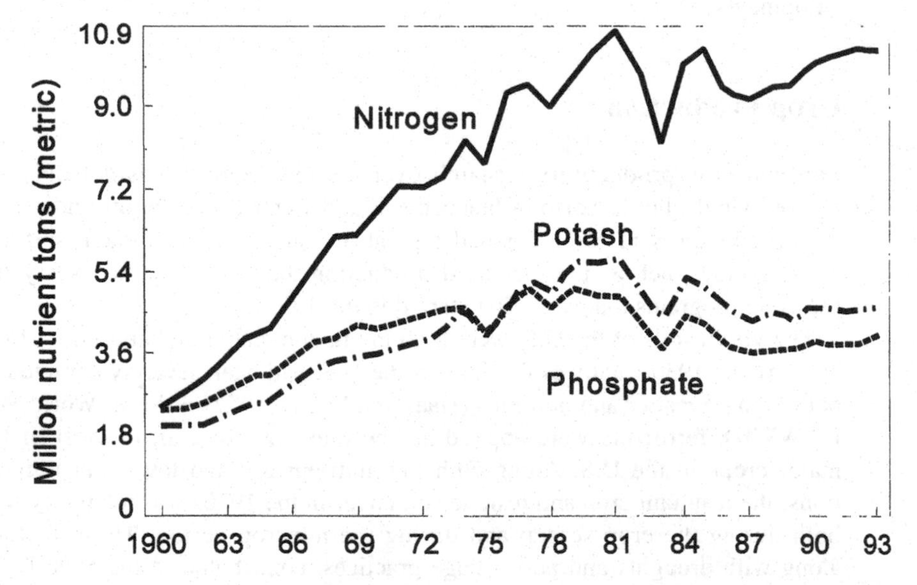

Figure 2. U.S. commercial fertilizer use, 1960–1993. (From Economic Research Service, 1994).

Research during the current century has increased crop production by several fold. The increased growth per unit area has benefitted protection of the soil from degradation by (i) greater production of residues for surface cover, (ii) increased canopy cover, (iii) larger amounts of residues for maintenance of soil organic matter, and (iv) the potential for taking lands vulnerable to degradation out of row crop production. The combined benefits of increased production have resulted in federal programs such as CRP, production set-asides, improved programs for runoff and erosion control, and maintenance of soil quality from return of biomass to the soil.

Several authors have argued that crop production should be maximized on our best lands, so that lands more susceptible to damage can be used for recreation, wildlife, wetlands, forestry, and plant diversity (Waggoner, 1994; Avery, 1994; Taff and Runge, 1987). Research for maximum crop production on the least vulnerable lands may, in effect, help conserve the vulnerable lands. While maximizing crop production on the least vulnerable lands is not stated as an objective for CRP, taking the vulnerable lands out of production is the primary objective. With current world grain reserves low, calls may increase for maximum economic crop production research on the good lands.

Environmental Quality

Publication of the book *Silent Spring* in 1962 (Carson, 1962) and the first Earth Day in 1970 ushered in an era in soil and water conservation that emphasized the need for greater recognition of environmental issues in agriculture. Some of the issues of concern included water quality, land application of municipal wastes, uses of crops and crop residues for production of energy, downstream sediment contamination, and pesticides and other chemicals in foods. The research community quickly accepted the responsibility for these issues. The American Society of Agronomy established the *Journal of Environmental Quality* in 1972. This journal is now considered one of the authoritative journals in its field as a source of original research, reviews, and comments.

The nationwide environmental movement in the 1970s forcibly brought to our attention the need for proper disposal of sewage wastes (Page and Chang, 1994). At that time it was apparent that land application was a viable alternative to incineration, landfills, ocean dumping, or other means of disposal/utilization. At a 1973 workshop sponsored by the Environmental Protection Agency (EPA), the U.S.DA, and the National Association of State Universities and Land Grant Colleges (NASULGC), the research needs on land utilization of sewage sludge was outlined (NASULGC, 1974). The research community quickly responded. For example, the number of land applications related technical publications increased from 2 in 1971 to 99 in 1986 (Page and Chang, 1994). In 1989, one-third of all municipal sewage sludge in the United States was land applied. Importantly, two decades of research has culminated in the promulgation of Standards for the Disposal and Utilization of Sewage Sludge.* This defines the pollutant loading boundary of land application (U.S. Environmental Protection Agency, 1993).

The energy crisis, brought on by the petroleum production slowdown and price regulation by the OPEC nations in the early 1970s created great interest in development of alternative sources of energy. Some suggested sources were ethanol production from grains, crop residues, and specialty crops. It was realized that large-scale use of production from the land could have a major bearing on soil and water conservation. A number of major research studies delineated where crop residues could be removed without unduly increasing erosion (Larson, 1979). While the need for energy sources other than petroleum has eased somewhat, the studies led the way in demonstrating techniques for assessing changes in land management on the environment.

Water Quality

Sediments in runoff water from soil erosion have long been recognized as the major pollutant of our nation's surface waters. However, during the late 1970s and 1980s, reports were increasingly frequent that both surface and groundwaters were being contaminated by agricultural chemicals, primarily nitrate and other fertilizer nutrients, pesticides, and animal wastes (CAST, 1985; U.S. Geological Survey, 1984; Madison and Burnett, 1985). More recently, concerns have been expressed about instances of contamination of drinking water supplies with such pathogenic organisms as *Cryptosporidium* and *Giardia*, with dairy cattle and other agricultural livestock suggested as potential sources (Flanagan, 1994). Major improvements in chemical analytical equipment and procedures have facilitated these findings by making possible the detection and identification of chemicals and other pollutants in water at levels of concentration many-fold less than were detectable before. The significance of such small levels of contaminants in water is still not clearly understood.

In response to these increasing concerns, the USDA initiated a national Water Quality Research Plan in 1989 (USDA, 1989) which was funded in 1990 by Congress through the President's Initiative on Water Quality. This Initiative is a multidisciplinary, multiagency program to address nonpoint source water quality problems related to agriculture. The USDA is the lead agency of the Water Quality Initiative, with strong input from the U.S. Geological Survey (USGS) and the EPA. Agencies within the USDA cooperating with active water quality programs include the ARS; CSREES; Economic Research Service (ERS); Consolidated Farm Service Agency (CFSA); Forest Service (FS); National Agricultural Library (NAL); National Agricultural Sta-

*Code of Federal Regulations, Title 40, Parts 257 and 503.

tistics Service (NASS); and the NRCS. The combined efforts of these agencies are coordinated and focused on the goals of the Water Quality Initiative by the USDA Working Group on Water Quality composed of representatives of each agency.

Research funding under the Water Quality Initiative was made available to the ARS and CSREES. Grants were made jointly by both the ARS and CSREES to establish five large-scale research projects in the Midwest to gain a better understanding of the impact of currently used agricultural management systems on water quality, and to develop agricultural production systems that are both economically profitable and environmentally beneficial. The Management Systems Evaluation Areas (MSEA) program was initially established at 10 research sites, which are representative of major groundwater aquifers and soil types in the Midwest, but has been expanded to include watershed-scale, systems research studies on water quality at other locations in the U.S. The MSEA program is an outstanding example of interdisciplinary research according to a 1995 external review, and addresses Congress' requirements for closer coordination among federal agencies. The MSEA research has successfully quantified loads of nutrients and major herbicides entering surface- and groundwater, and has shown a number of promising crop production practices that both are profitable and reduce nonpoint source water pollution, including modifications of tillage, herbicide, and nitrogen management (USDA, 1995).

Included in the USDA Water Quality Initiative nationwide are funds to conduct the CSREES Water Quality Special Research Grants Program. This program along with its counterpart "Water Resources Assessment and Protection" program in the National Competitive Research Grants Program, funds nationally competitive awards to conduct water quality research on such areas as: nitrogen testing, soil-water-chemical relationships and interactions, fate and transport of chemicals in soil and water, management practices and remediation practices to reduce water contaminants from agriculture, geographic information systems and decision aids to enhance applications of chemicals and water to lands, and socioeconomic and policy studies to enhance water quality.

Major partners with the federal government in water quality research are the State Agricultural Experiment Stations (SAES) and Land Grant Universities. The federal funds are magnified as much as four times or more by investments from state, university and local agencies to support this research focused on local and regional problems. All of these research areas are being closely coordinated with technology transfer programs in Extension and NRCS, as well as industry and agricultural producers to encourage the rapid adoption on the land. These cooperative federal-university research programs have already had impacts on policy by providing policy-makers the scientific research results on which to base federal, state, and local legislation and programs, such as the federal Clean Water Act, conservation titles in Farm Bill legislation, and state and local agricultural and environmental policies and programs.

Cropland Reductions

Starting in 1933, government acreage reduction programs have been an important part of U.S. cropland use. During that time, up to 28 million ha have been diverted to uses other than the production of major crops under government programs. Figure 3 summarizes the amounts of land set-asides under the variety of programs (ERS, 1994). Except for the Soil Bank Program of the late 1950 and 1960s and the current Conservation Reserve Program (CRP), the set-asides have all been short-term.

Adoption of soil conservation practices has been closely tied to government conservation payments. As early as the 1930s, payments were made for replacing planting of "soil-depleting" crops with planting "soil-conserving" crops. In general, soil-depleting crops were cash crops in surplus, and soil-conserving crops were perennial grasses and legumes. Payments were also made for installing conservation practices such as terracing and contouring. Government payments for these practices stimulated research as to the benefits from the practices and to improving their design.

As early as 1977, calls were made for giving priority attention to land and water most vulnerable to damage. Partly in response, the U.S. Congress passed the RCA Act of 1977. In response to this Act the USDA (headed by NRCS) developed the NRI, the results of which are available for 1982, 1987, and 1992. In these reports the extent of, and damages therefrom, have been estimated by comparing the calculated erosion using the USLE or WEQ with T. The CRP provisions of the 1985 Farm Bill provided a new approach to set-aside cropland reductions. The CRP was a giant step forward in that an attempt was made to target land based on its susceptibility for damage. Erosion was estimated by use of the research-developed U.S.LE (Wischmeier and Smith, 1978) and WEQ (Skidmore and Woodruff, 1968). Prior to CRP the choice of land for acreage reduction was given to the farmer. While many of the acreage reduction programs were termed conservation programs, they were primarily designed for production control.

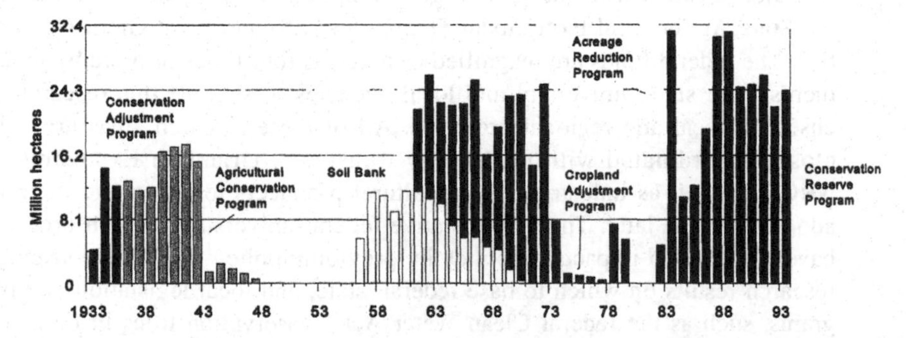

Figure 3. Cropland reductions by type of program 1944–1993. (From Economic Research Service, 1994).

The FSA, FACTA and FAIRA farm bills of 1985, 1990, and 1996 respectively, had provisions for highly erodible land (HEL) and wetland protection. Wetland protection is discussed in the wetlands section of this paper. Soil survey information is essential for the definition of HEL.* Data in SIR are used to calculate the soil erodibility index to determine if a map unit is HEL. These values are used to develop a highly erodible soil map unit list for each soil survey area (county) in the U.S. This list plus soil survey maps are used to determine if a field meets HEL requirements. The FSA, FACTA, and FAIRA contained provisions for the CRP, which has incentives for landowners/managers to return highly erodible land to permanent cover in return for a "rent payment" by the government. Soil survey information was used in the last three CRP sign-ups to evaluate rent bids submitted by the landowners and to rate the land offered with respect to water quality concerns. Average yields from the SIR for the predominant soil on the parcels bid into the program were used to evaluate the rent bids. Interpretations for potential pesticide and nutrient leaching hazards were used to rate the parcel with respect to potential water quality concerns.

In administering the RCA appraisals, the FSA of 1985, and the FACTA farm bill of 1990, the USLE and WEQ have been used to estimate erosion. They do not provide information on the distance of movement and location of deposition. Runge, Larson and associates (Taff and Runge, 1987; Roloff et al., 1988; Larson et al., 1988) have argued that in addition to designating land eligibility for CRP or similar programs by an estimate of the amount of erosion at a point, there should be a criterion for estimating the amount of potential productivity loss from a given erosion amount. Thus, they proposed a two-dimensional diagram as illustrated in Figure 4 which allows for both the potential resistivity of soil to damage and the loss in productivity. Indeed, this approach is used in the Reinvest in Minnesota program which is the state equivalent of CRP. Pierce (1987) has suggested that a three-dimensional diagram is needed in which resistivity, productivity, and environmental sensitivity all be considered.

Wetlands

The Wetland Reserve Program (WRP) was authorized by the FACTA and recently was transferred to the NRCS. Other wetland programs are administered by the U.S. Army Corps of Engineers (COE) and the EPA. A memorandum of agreement has been signed by the USDA, COE, EPA, and U.S. Fish and Wildlife Service (FWS) to develop and use consistent criteria and methods for delineating wetlands. An effort is underway to develop a hydrogeomorphic method (HGM) to estimate functions of wetland systems. Soil data and inter-

*Public Law 99-198 FSA, Public Law 101-624 FACTA, 7 CFR Part 12, NRCS staff, 1994.

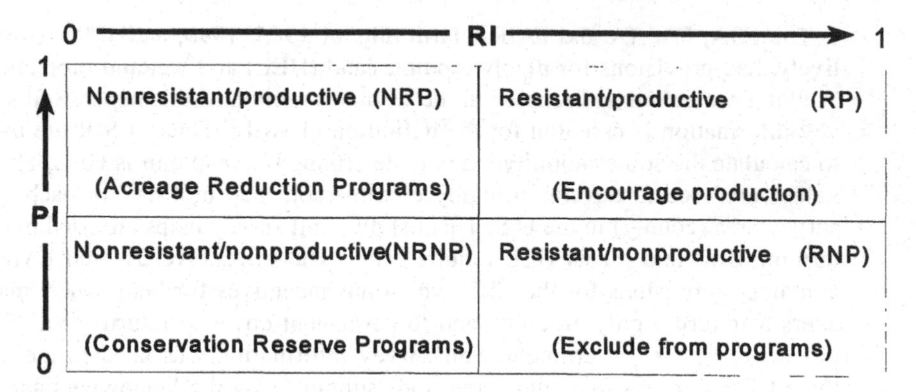

Figure 4. Locating soils in productivity/resistivity space (From Taff and Runge, 1987). PI refers to the Productivity Index and RI to the Resistivity Index (From Roloff et al., 1988).

pretations are critical factors in the HGM. Both swampbuster and the WRP require the identification of hydric soils as a part of the definitions of wetlands and that, in addition to required hydrology and hydrophytic vegetation, a parcel of land must have a predominance of hydric soils for it to be considered a wetland.

The term hydric soils was first coined by Cowardin et al. (1979) when they proposed that hydric soils be used as a criterion for identifying wetlands. Cowardin developed the nomenclature for wetland identification in preparation for the National Wetlands Inventory of the FWS. The FWS initially used soil survey information and definition of hydric soils to aid in this survey. Even though Cowardin introduced the term hydric soils, the Soil Conservation Service in conjunction with National Technical Committee for Hydric Soils (NTCHS), which included scientists from the Environmental Protection Agency (EPA), U.S. Corps of Engineers (COE), FWS, and universities developed the definition and criteria for hydric soils (Mausbach, 1994). The NTCHS first published the definition in 1985 (NTCHS, 1985).

Hydric soils are defined by criteria that use taxonomic classification, water table depths, flooding duration and frequency, and drainage class. These attributes are part of the SIR for all soil series of the U.S. The hydric soil criteria were developed to facilitate the development of a national list of hydric soils (Mausbach, 1994). However, the criteria for hydric soils are not easy to use in the field since they require wetland delineators to classify the soil and to measure depths to seasonal water tables. Water tables vary with time of year, and laboratory analyses are often required. Thus, soil scientists have recently developed a set of hydric soil indicators for use in the field. The hydric soil indicators are based on morphology linked to wetness and anaerobic conditions in soils. They allow experienced soil scientists to identify hydric soils in the field (Mausbach, 1994; Hurt and Puckett, 1993; Watts and Hurt, 1991).

Summary

While soil and water research was started in North America in colonial days, the first major United States Congressional action was passage of the Land Grant College Act in 1862. Other early acts were the Hatch Act establishing the Agricultural Experiment Stations in 1889, and the Smith-Lever Act of 1914 establishing the Extension Service.

In the 1920s erosion was recognized as a major threat to continued crop production. In response, Congress established the Soil Conservation Service (now the Natural Resources Conservation Service) within the USDA in 1935. In a reorganization in 1953, all soil survey activities were placed in the SCS and all research was placed in a new agency called the Agricultural Research Service. Forest research has continued in the Forest Service and economic research in the Economic Research Service within the USDA.

The soil survey program of the cooperative state and federal agencies has been the cornerstone for nearly all federal soil and water programs. Soil surveys are used in developing soil management and use programs for individual farmers and land owners, in the RCA Assessments, in Conservation Compliance, wetland assessment and management, and many other federal, state and local concerns.

Soil and water policy in the United States has closely followed research developments. The basis for erosion prediction on cropland was established in the 1930s and improvements continue today. Erosion prediction, initially developed as a guide for conservation planning, has been used in a host of government programs including the NRI and RCA Act, the CRP, the Sodbuster Act and many others.

Improved tillage and management practices have made the Conservation Compliance provisions of the 1990 Farm Bill feasible. Without modern conservation tillage systems, compliance would not be possible on many lands. Development of conservation tillage has been the primary factor in the recent national reductions in erosion amount. Production research has provided not only higher yields but has supplied more residues to protect the surface and for soil organic matter maintenance and improvement. The higher yields have also freed up marginal lands for other nondegrading uses including recreation and wildlife.

Soil and water research has also contributed significantly to both water quality and quantity policy programs. Since 1989, the federal government has provided over 30 million dollars in new research on water quality. These are now providing the basic data for best management practices for both surface and groundwater quality improvements.

Acronyms Used

ARS	Agricultural Research Service
BPISAE	Bureau of Plant Industry, Soils, and Agricultural Engineering

CFR	Code of Federal Regulations
CFSA	Consolidated Farm Services Agency
COE	U.S. Army Corps of Engineers
CRP	Conservation Reserve Program
CSREES	Cooperative State Research Education and Extension Service
EPA	Environmental Protection Agency
ERS	Economic Research Service
FACTA	Food, Agriculture, Conservation, and Trade Act of 1990
FAIRA	Federal, Agriculture, Improvement and Reform Act of 1996
FS	Forest Service
FSA	Food Security Act of 1985
FWS	Fish and Wildlife Service
HEL	Highly Erodible Land
LCC	Land Capability Class
NAL	National Agricultural Library (now part of ARS)
NASS	National Agricultural Statistics Service
NCSS	National Cooperative Soil Survey
NRCS	Natural Resources Conservation Service
NRI	National Resources Inventory
RCA	Resource Conservation Act of 1977
RUSLE	Revised Universal Soil Loss Equation
SAES	State Agricultural Experiment Station
SCS	Soil Conservation Service
SIR	Soil Interpretations Record
T	Soil Tolerance Value
USDA	United States Department of Agriculture
USGS	United States Geologic Survey
USLE	Universal Soil Loss Equation
WEPP	Wind Erosion Prediction Project
WEQ	Wind Erosion Equation
WRP	Wetland Reserve Program

References

Allmaras, R.R., S.M. Copeland, J.F. Power, and D.L. Tanaka. 1994. Conservation tillage systems in the northernmost central United States. In: Conservation Tillage in Temperate Agroecosystems. M.R. Carter (ed.) Lewis Publishers, Boca Raton, FL.

Allmaras, R.R., D.E. Wilkins, O.C. Burnside, and D.J. Mulla. 1998. Agricultural technology and adoption of conservation practices. pp. 99–158. In: F.J. Pierce and W.W. Frye (eds.) Advances in Soil and Water Conservation. Ann Arbor Press, Chelsea, MI.

Anderson, J.L., P.C. Robert, and R.H. Rust. 1992. Productivity and crop equivalent ratings for soils of Minnesota. AG-BU-2199-F Minnesota Extension Service University of Minnesota, St. Paul, MN.

Avery, D.T. 1994. Saving the planet with high–yield farming. Presented at Minnesota Plant Food and Chemical Association, Dec. 7, 1994. Minneapolis, MN.

Bartelli, L.J., A.A. Klingebiel, J.V. Baird, and M.R. Huddleson (eds). 1966. Soil surveys and land use planning. Soil Science Society of America. Madison, WI.

Bennett, H.H., and W.R. Chapline. 1928. Soil erosion: A national menace. USDA Circular No. 33. Washington, D.C.

Betts, E.M. 1953. Thomas Jefferson's Farm Book (with commentary and relevant extracts from other writings). American Philosophical Society and Princeton University Press. Princeton, NJ.

Blevins, R.L., R. Lal, J.W. Doran, G.W. Langdale, and W.W. Frye. 1998. Conservation tillage erosion control and soil quality. pp. 51–68. In: F.J. Pierce and W.W. Frye (eds.) Advances in Soil and Water Conservation. Ann Arbor Press, Chelsea, MI.

Browning, G.M., C.L. Parish, and J.A. Glass. 1947. A method for determining the use and limitation of rotation and conservation practices in control of erosion in Iowa. Soil Sci. Soc. Am. Proc. 23:246–249.

Carson, Rachel. 1962. Silent Spring. Houghton Mifflin, Boston.

Clark, E.H. II, J.H. Haverkamp, and W. Chapman. 1985. Eroding soils: The off–farm impacts. The Conservation Foundation, Washington, D.C.

Colacicco, D., T. Osborn, and K. Ault. 1989. Economic damage from soil erosion. J. Soil Water Conserv. 44:35–39.

Cooper, D.T., and R.V. Bollman. 1989. Reaching for soil conservation's ultimate goal in Indiana. J. Soil Water Conserv. 44:392–394.

Council on Agricultural Research and Technology (CAST). 1985. Agriculture and Ground Water Quality. CAST Report No. 103. Ames, IA.

Cowardin, L.M., V. Carter, F.C. Golet, and E.T. LaRoe. 1979. Classification of wetlands and deepwater habitats of the United States. Fish and Wildlife Service, U.S. Dept. of the Interior, Washington, D.C.

Crosson, P. and N.J. Rosenberg. 1989. Strategies for Agriculture. Scientific American 261:128–137.

Crosson, P. 1986. Soil erosion and policy issues. In: T. Phipps, P. Crosson, and K. Price (eds.), Agriculture and the Environment. Resources for the Future, Washington, D.C.

Doran, J.W., D.C. Coleman, D.F. Bezdicek, and B.A. Stewart (eds.) 1994. Defining Soil Quality for a Sustainable Environment. Soil Science Society of America, Special Publication No. 35. Soil Science Society of America, American Society of Agronomy. Madison, WI.

Duley, F.L., and J.C. Russell. 1942. Machinery requirements for farming through crop residues. Agric. Eng. 23:39–42.

Duley, F.L., and J.C. Russell. 1939. Annual Research Report. United States Department of Agriculture, Lincoln, NE.

Economic Research Service. 1994. Agricultural resources and environmental indicators. U.S. Department of Agriculture. Washington, D.C.

Flanigan, T.P. 1994. A massive outbreak in Milwaukee of *Cryptosporidium* infection transmitted through the public water supply. New England J. Med. 331:161–167.

Fryrear, D.W. and J.D. Bilbro. 1998. Mechanics, modeling, and controlling soil erosion by wind. pp. 39–49. In: F.J. Pierce and W.W. Frye (eds.) Advances in Soil and Water Conservation. Ann Arbor Press, Chelsea, MI.

Gray, L.C. 1933. History of agriculture in the southern United States to 1960. Carnegie Institution of Washington, Washington, D.C.

Haas, H.J. and C.E. Evans. 1957. Nitrogen and carbon changes in Great Plains soils as influenced by cropping and soil treatments. U.S.DA Tech. Bull. No. 1146. Washington, D.C.

Hurt, G.W., and W.E. Puckett. 1993. Proposed hydric soil criteria and their field identification. In: Proceedings of the Eighth International Soil Management Workshop: Utilization of Soil Survey Information for Sustainable Land Use. J.M. Kimble (ed.) USDA Soil Conservation Service, National Soil Survey Center. May 1993. pp. 148–151.

Johnson, L.C. 1987. Soil loss tolerance: Fact or myth. J. Soil Water Conserv. 42:155–160.

Klingebiel, A.A. 1991. Development of soil survey interpretations. Soil Surv. Horizons. 31:53–66.

Laflen, J.M. 1998. Understanding and controlling soil erosion by rainfall. pp. 1–19. In: F.J. Pierce and W.W. Frye (eds.) Advances in Soil and Water Conservation. Ann Arbor Press, Chelsea, MI.

Larson, W.E., F.W. Schaller, W.G. Lovely, and W.F. Buchele. 1956. New ways to prepare corn ground. Iowa Farm Sci. 10:3–6.

Larson, G.A., G. Roloff and W.E. Larson. 1988. A new approach to marginal agricultural land classification. J. Soil Water Conserv. 43:103–106.

Larson, W.E., C.E. Clapp, W.H. Pierre, and Y.B. Morachan. 1972. Effects of increasing amounts of organic residues on continuous corn: II Organic carbon, nitrogen, phosphorus, and sulfur. Agron. J. 64:204–208.

Larson, W.E. 1979. Crop residues: Energy production or erosion control. Soil Conservation Society of America, Special Publications No. 25. Ankeny, IA.

Lowdermilk, W.C. 1953. Conquest of the land through 7000 years. Bulletin 99. U.S. Department of Agriculture, Soil Conservation Service, Washington, D.C.

Madison, R.J. and J.O. Burnett. 1985. Overview of the occurrence of nitrate in ground water of the United States. National Water Summary 1984-Hydrologic.

Mausbach, M.J. 1994. Classification of wetland soils for wetland identification. Soil Surv. Horizons 35:17–25.

Middleton, H.E., C.S. Slater, and H.G. Byers. 1932. Physical and chemical characteristics of the soils from the erosion experiment stations. U.S. Dept. Agr., Tech. Bull. No. 316. Washington, D.C.

Musgrave, G.W. 1947. The quantitative evaluation of factors in water erosion, a first approximation. J. Soil Water Conserv. 2:133–138.

National Association of State Universities and Land Grant Colleges. 1974. Proceedings of the Joint Conference on Recycling Municipal Sludges and Effluents on Land. Champaign, IL., 9–13 July 1973. NASULGC, Washington D.C.

National Research Council. 1993. Soil and water quality: An agenda for agriculture. National Academy Press, 2001 Constitution Avenue, Washington, D.C.

National Technical Committee for Hydric Soils. 1985. Hydric soils of the United States. USDA Soil Conservation Service, Washington, D.C.

Obenshain, S.S. 1966. Changes in the need and use of soils information. In: Soil Surveys and Land Use, pp. 175–179. L.J. Bartelli, A.A. Klingebiel, J.V. Baird, and M.R. Huddleston (eds.) Soil Science Society of America, Madison, WI.

Page, A.L. and A.C. Chang. 1994. Overview of the past 25 years: Technical perspectives. In: Sewage Sludge: Land Utilization and the Environment, C.E. Clapp, W.E. Larson, and R.H. Dowdy (eds.) Soil Science Society of America, Madison, WI.

Pierce, F.J., R.H. Dowdy, W.E. Larson, and W.H.P. Graham. 1984. Soil productivity in the Corn Belt: An assessment of erosion's long-term effects. J. Soil Water Conserv. 39:131–138.

Pierce, F.J. 1987. Complexity of the landscape. In: Making Soil and Water Conservation Work: Scientific and Policy Perspectives. D.W. Halbach, C.F. Runge and W.E. Larson (eds.) Soil and Water Conservation Society of America, Ankeny, IA.

Pierce, F.J. and E.J. Sadler, ed. 1997. The state of site-specific management for agriculture. ASA Misc. Publ. ASA, CSSA, and SSSA, Madison, WI.

Renard, K.G., G.W. Foster, G.A. Weesies, and J.P. Porter. 1991. RU.S.LE: Revised Universal Soil Loss Equation. J. Soil Water Conserv. 46:3033.

Roloff, G., G.A. Larson, W.E. Larson, R.J. Voss, and W.W. Becken. 1988. A dual targeting criterion for soil conservation programs in Minnesota. J. Soil Water Conserv. 43:99–102.

Shellhorn, Gary S. 1993. Integration of soil information for land use planning in the Lake Tahoe Region: Land Classification as a runoff and erosion control technique. In: Proceedings of the Eighth International Soil Management Workshop: Utilization of Soil Survey Information for Sustainable Land Use, pp. 230–235. J.M. Kimble (ed.) USDA Soil Conservation Service, National Soil Survey Center. May 1993.

Shertz, D.L. 1983. The basis for soil loss tolerances. J. Soil Water Conserv. 38:10–14.

Simonson, Roy W. 1987. Historical aspects of soil survey and soil classification. Soil Science Society of America, Madison, WI.

Skidmore, E.J. and N.P. Woodruff. 1968. Wind erosion forces in the United States and their use in predicting soil loss. Agricultural Handbook 346. U.S. Department of Agriculture, Washington, D.C.

Smith, D.D. and D.W. Whitt. 1947. Estimating soil losses from field areas of claypan soils. Soil Sci. Soc. Am. Proc. 12:485–490.

Soil Conservation Service Staff. 1989. The Second RCA Appraisal. Soil, water, and related resources on nonfederal land in the United States, analysis and trends. USDA Miscellaneous Publication Number 1482. Washington, D.C.

Soil Conservation Service Staff. 1994. National Food Security Act Manual, 3rd edition. USDA, Soil Conservation Service, Washington, D.C.

Soil Survey Staff. 1992. Soil Survey Division Program Plan. USDA Soil Conservation Service, Washington, D.C.

Taff, S.J., and C.F. Runge. 1987. Supply control, conservation, and budget restraint: Conflicting instruments in the 1985 Farm Bill. In: Making Soil and Water Conservation Work: Scientific and Policy Perspectives, D.W. Halbach, C.F. Runge and W.E. Larson (eds.) Soil and Water Conservation Society. Ankeny, IA.

U.S. Department of Agriculture, Agricutlural Research Service. 1995. Farming systems impact on water quality. Management Systems Evaluation Areas (MSEA). Progress Report, 1994 (in press).

U.S. Department of Agriculture. 1989. USDA research plan for water quality. U.S.DA Agricultural Research Service and Cooperative State Research Service, Washington, D.C. January 1989.

U.S. Environmental Protection Agency. 1993. Standards for the use or disposal of sewage sludge. Federal Register 58(32):9248–9415. U.S. Printing Office, Washington, D.C.

U.S. Geologic Survey. 1984. National Water Summary. 1984. Hydrologic Events, Selected Water Quality Trends, and Ground Water Resources. U.S. Geological Survey Supply Paper 2275. Department of the Interior, Washington, D.C.

Watts, F.C., and G.W. Hurt. 1991. Determining depths to the seasonal high water table and hydric soils in Florida. Soil Surv. Horizons 32:117–121.

Waggoner, P.E. 1994. How much land can ten billion people spare for nature? Council for Agricultural Science and Technology, Report 121. Ames, IA.

Wischmeier, W.H. and D.D. Smith. 1965. Predicting rainfall-erosion losses from cropland east of the Rocky Mountains—Guide for selection of practices for soil and water conservation. USDA Agricultural Handbook No. 282. Washington, D.C.

Wischmeier, W.H., and D.D. Smith. 1978. Predicting rainfall erosion losses. Agr. Handbook No. 537. U.S. Dept. Agr., Washington, D.C.

Zingg, R.W. 1940. Degree and length of land slope as it affects soil loss in runoff. Agric. Eng. 21:59–64.

Soil and Water Conservation: Soil Science in the Next Half-Century

T.C. Daniel, W.D. Kemper, and J.L. Lemunyon

Introduction

In the early 1930s, the visionary J.N. "Ding" Darling began to put forth ideas and concepts about conservation that are topics for today's researchers, conservationists and agencies (Darling, 1961). Mr. Darling, employed by the *Des Moines Register*, published his innovative ideas on conservation as a syndicated cartoonist. His simple drawings distilled complex issues into concepts easily understood by the public. Cast in today's jargon, he employed "user-friendly" techniques to communicate effectively with a known clientele. He understood the relationship between land use and water quality (Figure 1) and that natural practices (Figure 2) on the land—precursors to today's best management practices—would optimize the impact of human activity. His work was widely circulated by the national press, elevating the public's consciousness about conservation issues and contributing to passage of national legislation on soil and water conservation.

This chapter attempts to visualize the issues and challenges to be encountered during the next 50 years and, perhaps most importantly, to identify broad strategies to assist soil scientists in meeting such needs. We can count our efforts successful if we are one-tenth as accurate in our projections as Mr. Darling. Thanks, "Ding."

What Do We Know About the Future?

Through time, every generation has been subjected to unique circumstances that influenced future events. To increase the likelihood of success in visualizing the future, it is important to recognize some of today's circumstances that will mold the role of the soil scientist during the next half-century. The following is a brief profile of some important factors accompanied by projections as to their ramifications.

Communication

The first fax (facsimile) was transmitted from Lyon to Paris in 1865, but only recently has this technology been available to the general public due to lower

AS LAND GOES SO GOES MAN

BUT JUST TRY TO GET ANY POLITICIAN TO DO ANYTHING ABOUT IT

Figure 1. Effect of land use on erosion and water quality. From J.N. "Ding" Darling, 1961 (plate 32). Reprinted with permission by The Des Moines Register and Tribune Company, 1961.

transmission costs (Swerdlow, 1995). The fax is only a pixel of the "information revolution" because technology now exists that will foster future revolutionary breakthroughs. If you have marveled recently at the capability of the latest computer or the Internet, then prepare yourself for some real jolts from

Figure 2. Nature's best management practice. From J.N. "Ding" Darling, 1961 (plate 37). Reprinted with permission by The Des Moines Register and Tribune Company, 1961.

cyberspace. For example, one current CD-ROM disk holds the information contained on 330,000 sheets of single-spaced text. The next generation of disks will likely be able to put a full public library at your fingertips (Swerdlow, 1995). Wireless handheld receivers (visual and sound) will be available to the public with links to e-mail, Intenet, fax, satellites, and voice communication.

One might ask "What does the information superhighway have to do with the role of soil science in 2050?" Some possibilities: Think of the valuable information available to farmers interested in conservation tillage if they have instant access from the kitchen table or mobile office (pickup) to technical information from industries, research specialists and, perhaps most importantly, other farmers already practicing the techniques in different regions of the country. State and regional boundaries will dissolve due to the ability to communicate electronically. At the research level, effective meetings will be held via satellite where scientists from different disciplines will collaborate on an interdisciplinary paper or assist an agency to identify priority watersheds to receive scarce resources. As outlined by Miller (1993), Gardner (1993), and Warkentin (1994), most future problems will be best solved through an interdisciplinary team approach. The ability to communicate freely and effortlessly, regardless of location, is the first step in implementation.

Because of the information revolution, the urban public will be capable of accessing heretofore unavailable information for decision making. For example, in Swerdlow's (1995) excellent article on the information revolution, an account is given of an environmental problem associated with a chemical plant. A local individual was able to access the Right-to-Know Network, determine that toxic chemicals were being emitted from the plant, and verify that the emissions likely accounted for the flu-like symptoms she and her neighbors were experiencing. The information revolution will have a very direct impact on the environment. For example, the ability to store information contained in a library on a single disk will have significant implications on the paper industry and forestry management. And the burden on available landfill space should lessen as a result of this technology, especially as electronic newspapers become available.

Demographics

By the year 2050, the world's population is estimated to be around 10 billion (Mullins, 1993). During this same period, the U.S. population is expected to grow from around 260 to 375 million (Bremner and Weber, 1992), and the number of people over the age of 65 is expected to nearly double (Stacey, 1995). American diets will change too. Today, over 50% of the household food expenditure is for meals prepared outside the home. Shortages in meat and dairy products and changes in eating habits will move the consumer to more grains and beans. The general population will continue to change from predominantly agrarian to urban. Most future soil scientists will be born and reared in urban environments. For these scientists, farm and ranch experience will be limited to brief contacts with producers. The context of soil investigations will likewise be less from the farm and ranch setting and more from urban and nonfarm areas.

Demand on the Resource

The shift to an urban society will have profound effects on soil and water conservation and the role played by the soil scientist. Simply put, the urban public, with its interest in environmental concerns, will have the political clout to influence the legislative process to the extent that its wishes become reality. For example, in 1992, water rights became an issue between urban and agricultural interests in the Central Valley of California. A coalition was formed between business leaders, urban water agencies, and environmentalists to break agriculture's monopoly on Central Valley Project water (Conniff, 1993). The logic was simple. Of the 40 million people in California, there were more urban than agricultural votes, so in the reallocation, agriculture was forced to relinquish some of its traditional rights to water and consider water marketing as an alternative to water use.

The urban public will demand water of high quality for drinking and recreation. For example, spring runoff from corn and soybean fields in the Midwest transport herbicides to the Mississippi River via its many tributaries (Goolsby et al., 1991; Ellis, 1993). Many municipalities, such as the city of New Orleans, use the river water as a drinking water supply, and herbicide concentrations often (usually coinciding with spring application and runoff) exceed the maximum level deemed advisable by the U.S. Environmental Protection Agency. In the future, the urban population can be expected to be better informed and less tolerant of contaminants in its drinking water and will force better management of agricultural nonpoint sources of pollution. It will also demand a cheap supply of readily available high quality food that is produced in an environmentally friendly manner. Aquatic resources and landscapes will continue to exhibit stress from overuse by more people with more free time. Some go so far as to predict environmental doom should the population double by 2050 (Linden, 1992). The U.S. is a big user of water, three times more per capita than the average European country and far more yet than most developing nations.

Mega-volumes of organic wastes will be produced by both the human population during the conduct of its daily activities and the animal population confined in feedlots to streamline protein production. Innovative recyclable avenues that utilize such wastes in an environmentally friendly manner beg for invention; until then, however, the major use of such materials will be through sustainable agriculture. However, the public will have more information on the potential hazards of wastes and will require that their contents be analyzed to allow exclusion of materials bearing potentially hazardous levels of components such as heavy metals. A realistic assessment of the hazards involved will be needed to achieve the optimum balance between recycling of wastes to sustain our production and assuring the public a safe supply of food. Concerns regarding pathogens, such as *Cryptosporidium*, will result in a demand for treatment systems that produce a sanitized product, e.g., composting or methods involving heating or chemical treatment. The relatively high costs of such treatments will contribute to a demand for information on the ability of specific soils to remove pathogens from the water.

Clientele Shifts

To date, a mutually beneficial symbiotic relationship has existed between the land-grant universities, the USDA, farmers, and production agriculture. In the context of an effective framework for information exchange, this relationship has worked well because each group knew its clientele and the role it needed to play. Now, we are in a transition stage trying to define who we are, what we do best, and who needs our services (*deja vu* '60s discussion). As the population becomes more urban, a smaller portion of our potential clientele will be involved in production agriculture, and opportunities for service with the urban public will be a larger part of the mission of future soil scientists. What eventual form this mission takes is anyone's guess, but it certainly will be oriented toward working with and providing answers to concerns of the urban public. To bring this to fruition, the soil science profession needs objective spokespersons who understand basic principles, who are able to define the potentials for improvement in our current systems, and who can realistically spell out where soil scientists are needed and their role, and describe how they will complement the contributions of other scientists and managers essential to achievement of those potentials.

The urban public has a strong interest in maintaining and improving the quality of their environment, including the quality of their air, water, gardens, parklands, and recreational areas to which soil scientists can contribute. For example, results from a recent survey conducted by the USDA-Natural Resources Conservation Service (NRCS) show that the general public feels that water quality, air pollution, and solid waste management are the most crucial natural resource problems in their own communities. Yet, nearly one-fourth of the public responded that their community was not experiencing any natural resource problems (USDA-NRCS, 1995). A subgroup of the public, traditional soil and water conservation partners, responded that water quality, soil erosion, and agricultural sustainability were the most important natural resource issues in the next decade (USDA-NRCS, 1995). From both the general public's view and the traditional conservation partner's prospective, there are still issues needing the technical guidance of soil scientists. Both groups feel that farmers, ranchers, and homeowners are good caretakers of the resources. Industry, however, is considered a poor steward of its resources.

As discussed by Miller (1993), soil scientists need to seek an identity outside of agriculture if soil science is to be competitive with other professions in the future. Soil scientists need to be associated with organizations and organizational names that clearly indicate the relevance of soil science to urban as well as agricultural concerns. Will soil scientists have definable research needs and the money and infrastructure to fund research and disseminate its findings appropriately? While the private sector, both industry and citizen groups such as the Sierra Club, will continue modest funding to answer specific issues, urban concerns and their representatives will continue to be identified through the legislative process. If urbanites do not want pesticides in their drinking water, then legislative pressure will be exerted to ensure that appropriate agencies are identified and given the clout and money to act in their interest. For these reasons, then, state and federal agencies are promising clientele who need the profession's expertise and experience, especially as it relates to agricultural nonpoint pollution. At the federal level this would include the US-EPA, USDA-NRCS, and USDA-Agricultural Research Service (ARS). The state level would include the appropriate federal agencies plus those state agencies given the responsibility for protecting and preserving the environment.

Ramifications

What does all this mean in terms of possible impact? How will it change or impact recent graduates in soil science? If the preceding discussion has any validity, we cannot keep doing the same things if we wish to contribute in meaningful and professional ways to the future of agriculture. The following is our best guess.

- **Interdisciplinary.** Team efforts from various disciplines will be used to solve problems and understand processes.
- **Urban Clientele.** Interdisciplinary teams will focus on serving urban interests as represented by public agencies.
- **Sustainable Agriculture.** More food will be produced on less land by fewer farmers, resulting in less damage to the resource and better protection of the environment.
- **Environment.** Subject area for the future with efforts devoted toward maintaining and enhancing the quality of the air, water, and soil resources.
- **Communication.** The ability to keep current in the information revolution (hardware and technology) will be an inherent part of the professional soil scientist.
- **Outreach.** Information programs and materials will focus on urban concerns, especially the environment. Educational materials will be packaged and transmitted through in-place electronic networks.
- **Education.** Colleges and universities will emphasize the basics and the ability to communicate. Degrees will be awarded from regional land-grant universities (universities without borders) and more courses will be taken electronically, both undergraduate and graduate.
- **Information.** Information will be available in expanded detail. Site-specific data for soil, weather, management, and economics will enable

the soil scientist to perform site-specific tasks in a timely and efficient manner.

Soil Science Contribution

It is beyond our ability to predict specific technical contributions that will be made during the next half-century. Soil scientists will treat resource concerns (eroding soils, polluted air, contaminated water) more in respect to the public concern than to the landowner or traditional (farmers and ranchers) client's concern. Everyone feels ownership of our natural resources. Along with private property rights comes the responsibility to manage land resources in a way that does not degrade the "public's" environment. Technology of resource management, whether voluntarily applied or mandated, will be an important role of the soil scientist.

Technology, such as the use of polymers for erosion control (Lentz and Sojka, 1994; Laflen, 1996; Sojka, 1996), will continue to be developed, resulting in significant advancements in integrating environmental protection and sustaining agriculture. Such breakthroughs in environmental quality will be touted at the Soil and Water Conservation Society's centennial celebration as were advancements in production agriculture at the half-century. The following are broad research areas in soil and water conservation to which soil scientists will make significant contributions.

- **Interdisciplinary Research - Team Building.** Solving today's and tomorrow's problems cannot be done by a single discipline. As suggested by Gardner (1993), interdisciplinary teams will be emphasized and soil scientists will be key players.
- **Applied Research.** Funding agencies will emphasize research that produces solutions to real-world problems. Much of the traditional soil science research will be used for new applications. Some examples might include soil testing procedures to determine environmental thresholds, innovative measurements, such as soil redox potential, to identify wetlands, and improved methods of tillage for residue/biosolids management.
- **Water Use Efficiency.** Present-day competition between the urban and agricultural need for the same water will foster continued need for efficient use. Soil scientists will be needed to develop new methods of water conservation as well as technology to achieve more efficient use of our water.
- **Nonpoint Pollution from Agriculture.** While soil scientists will contribute to solving problems relating to urban nonpoint sources, the major contribution will occur in the agricultural nonpoint area.

- **Conservation Tillage.** Interest in conservation tillage will continue from an environmental as well as a production standpoint. Soil scientists will play significant roles in developing innovative equipment to circumvent problems such as low soil temperature in no-till, crop residue systems that foster the use of manure, and herbicide/fertilizer placement.
- **Nutrient Management.** Interest in nutrient management will increase as a means of maintaining water quality and attaining sustainable agriculture. Soil scientists will be involved in developing the technical details of sound nutrient management plans as well as user-friendly software.
- **Water Quality Criteria.** Soil scientists will interact with disciplines such as limnology and stream ecology to identify the impact that agricultural nonpoint pollution is having on water quality and determine how it can be circumvented.
- **Best Management Practices.** Management practices that protect and enhance resources while increasing production will require continued development and refinement. Innovative amendments influencing the chemical and physical characteristics of the soil will be developed.
- **Watershed Emphasis.** Research and water quality implementation programs will emphasize the holistic watershed management approach on a large scale. Soil scientists will be involved in developing methods to prioritize watersheds and identify (target) areas in the watershed that should first receive implementation of best management practices.
- **Manure and Biosolids.** Innovative ways of using, processing, and incorporating manure and biosolids into the agricultural and urban environment will be developed by soil scientists. In regions such as the Chesapeake Bay, where confined animal operations exist in close proximity to high-value aquatic resources, soil scientists will provide leadership in developing a manure hauling industry that can efficiently transport manure from where it is being overproduced to where it is viewed as a valuable resource for sustainable agriculture and improvement in soil quality.
- **Impact and Risk Assessment.** Soil scientists will work in the unaccustomed field of impact and risk assessment. For example, they will work with limnologists and stream ecologists in specific aquatic resources to assess the level of phosphorus loading from agriculture that produces realistic and acceptable levels of eutrophication. They will work with policy makers to establish concentration standards for pesticides in surface and groundwater that properly weigh the need for the compound and its risk to the general public.
- **Model Development and Use.** Models will be developed and used to reduce erosion, increase agricultural production and minimize the water quality impacts from agricultural nonpoint sources.

- **Information Transfer.** To gain credibility and acceptance from our clientele, information will be developed, packaged, and disseminated to appropriate clientele in an easily understood and time-efficient manner. To establish soil science as a profession on equal footing with other professions, this may be our most important and demanding area.

It is beyond the scope of this chapter to explore each of these areas in detail. However, as an example of the broad array of potentials for improvement in most of them, consider the following potential that falls primarily under **"Water Use Efficiency."**

Increasing needs of cities, industry, and recreation for water and their ability to pay higher prices than agriculture for our "collected" water will diminish water available for irrigation. Coupled with food, fiber, and energy needs of an increasing population, this demand on water resources will provide strong economic incentives for soil managers and scientists to use precipitation more efficiently where it falls. Water deficiency is the most common factor causing crop failures in the U.S., particularly on soils with restrictive horizons (clay pans, fragipans, compaction pans, acid subsoils) which limit effective rooting depths [<24 in. (60 cm)]. Crop yield on soils with restrictive horizons is highly dependent on timely rains, and yields currently exceed production costs in only about 50% of the years. Crop production models (Sinclair, 1994) indicate that drought stress exerted by these soils can be reduced sufficiently to enable market value of the crop to exceed production costs in 95% of growing years by doubling the amount of available water held by these soils.

Are such changes possible? Soil scientists have achieved increased rooting depth of this magnitude by using gypsum to increase the Ca content of acid subsoils and have improved root penetration in clay soils by deep chiseling during the previous fall when the soils are dry (Sumner, 1994; Wesley and Smith, 1991). Recent studies (E.E. Alberts, 1995, personal communication) on rooting depth of eastern gamagrass [*Tripsacum dactyloides* (L.)L.] found roots to depths of 59 to 83 in. (150 to 210 cm) in claypan soils where effective soybean and corn roots are commonly restricted to the top 24 in. (60 cm). Roots of soybean grown on soil with a long prior tenure in eastern gamagrass were found to a depth of 59 in. (150 cm). Aerenchyma in the roots of the perennial eastern gamagrass are apparently enabling them to obtain O_2 and survive the wet and often saturated conditions when root resistance of the tight clay soils is lowest. Evidence is accumulating to indicate that selected soybean cultivars have the ability to develop aerenchyma and thereby survive in flooded soils. All these advances in the soil/plant system could expand the reservoir of water available to the crops as a result of deeper rooting in wet soils. This mounting evidence that such improvements are possible provides soil managers and scientists with the exciting and potentially profitable challenge of devising systems in which these improvements are economically feasible for the broad spectrums of soil, climate, and crop characteristics.

There are some tendencies for managers to conclude that we know most of the basic facts needed for soil management and that all that remains is to properly formulate these facts into models and apply them to systems. However, there is a continuing need to evaluate our past conclusions in the light of new evidence. For instance, we have generally assumed that the available water holding capacity is the difference between the field capacity and the wilting point. However, if some species or varieties have the ability to survive and even thrive under saturated conditions in the soil, this would increase the "available water" holding capacity of these soils. Can we reduce the artificial drainage needed in our fields? Can roots of perennial species with aerenchyma penetrate enough to serve as a safety net to capture nitrate that has leached past the normal rooting depth of most of our crops? Such opportunities for soil scientists to cooperate with plant scientists in ascertaining how plant characteristics can reduce stresses limiting production while improving the environment is potentially one of our most productive areas.

These potentials for improvement and others, such as the ability of cover crops and no-till systems to increase the organic matter content of our soils, are clearly indicating that understanding processes leads to production of the needed food, fiber, and energy while improving soil and water resources. In the past, we have tended to accept conservation of these resources as our ultimate goals. However, the increasing demands for water, food, fiber, and energy associated with the world's growing population require that we enhance these resources and their abilities to satisfy those needs. Scientists working in close cooperation with soil managers can achieve the understanding of the processes necessary to make the improvements possible, and to link them synergistically into systems where they will be economically feasible and environmentally beneficial.

Summary and Conclusions

Demographic factors will have a profound influence on the discipline of soil and water conservation and on the role of soil scientists in the next 50 years. Urban population shifts will change our clientele and emphasize protecting and preserving environmental quality. Soil scientists will work with other disciplines in solving real-world problems. Communication technology will continue to advance the field of soil and water conservation, and professional soil scientists will be on the cutting-edge of this important technology. Sources and amounts of funding will continue to change but will maintain an applied emphasis. Past agriculture production technology will be used by soil scientists to produce food and fiber in an environmentally friendly manner. Soil scientists will continue to provide leadership in soil and water conservation. The clients will change along with the scope of the environmental concerns, the science will become more site-specific, and technology to measure and monitor the soil will advance very rapidly. However, the world's requirements and dependence on the soil resource will become crucial, and soil scientists will be there to sustain and enhance our natural resources.

If soil research budgets continue to diminish, there will be a tendency for some to conclude that society cannot afford to sustain its investment in soil research. Increasingly, soil scientists must assume the responsibility to plan and conduct research in cooperation with potential users in order to meet their needs and provide valuable products. In the past, we have relied heavily on decision makers to recognize the quality of our research products and support us appropriately. Pioneering research on soil fertility and water management has paid obvious dividends in increased production and helped develop the most cost-efficient food and fiber production systems in the world. Those pioneers gave us credibility and value in the food and fiber production arena but, while valuable, soil scientists must also establish their value in environmental, energy, and other arenas to merit society's continued support, particularly when the granaries are full.

References

Alberts, E.E. 1995. Personal communication. USDA-ARS. Columbia, MO.

Bremner, B., and J. Weber. 1992. A spicier stew in the melting pot. Business Week, 21 Dec., p. 29.

Conniff, R. 1993. California: Desert in disguise. In: W. Graves (ed.) Water—The power, promise, and turmoil of North America's fresh water. National Geographic Special Edition 184 (5A):38–53.

Darling, J.N. 1961. J.N. "Ding" Darling's conservation and wildlife cartoons. The J.N. "Ding" Darling Foundation, Inc. P.O. Box 703, Des Moines, IA 50303-0703.

Ellis, W. S. 1993. The Mississippi: River under siege. In: W. Graves (ed.) Water—The power, promise, and turmoil of North American's fresh water. National Geographic Special Edition 184 (5A):90–119.

Gardner, W.R. 1993. A call to action. Soil Sci. Soc. Am. J. 57:1403–1405.

Goolsby, D.A., R.C. Coupe, and D.J. Markovchick. 1991. Distribution of selected herbicides and nitrate in the Mississippi River and its major tributaries, April through June 1991. U.S. Dept. of the Interior, U.S. Geological Survey, Water Resources Div. Report (#91-4163). Denver, CO. p. 31.

Laflen, J.M. 1997. Scientific contributions to understanding and control of soil erosion by rainfall. In: F.J. Pierce and W.W. Fry (ed.) Research contributions to fifty years of soil and water conservation in the United States. Soil and Water Conservation Society of America, Ankeny, IA.

Lentz, R.D., and R.E. Sojka. 1994. Field results using polyacrylamide to manage furrow erosion and infiltration. Soil Sci. 158:274–282.

Linden, E. 1992. Too many people. Time 140 (27):64–65.

Miller, F.P. 1993. Soil science: A scope broader than its identity. Soil Sci. Soc. Am. J. 57:299, 564.

Mullins, M.E. 1993. As the world grows. USA Today. 4 June, Sec. A, p. 1.

Sinclair, T.R. 1994. Limits to crop yield? pp. 509–532. In: K. J. Boote, J.M. Bennett, T.R. Sinclair, and G.M. Paulsen (eds.). Physiology and determination of crop yield. ASA, CSSA, and SSSA. Madison, WI.

Sojka, R.E. 1996. Research contributions to the understanding and management of irrigation–induced erosion. In: F.J. Pierce and W.W. Fry (ed.) Research contributions to fifty years of soil and water conservation in the United States. Soil and Water Conservation Society of America, Ankeny, IA.

Stacey, J. 1995. Graying of the USA. USA Today. 8 May, Sec. A, p. 1.

Sumner, M.E. 1994. Amelioration of subsoil acidity with minimum disturbance. pp. 231–240. In: N.S. Jayawardane and B.A. Stewart (ed.). Subsoil Management Techniques. Lewis Publishers, Boca Raton, FL.

Swerdlow, J.L. 1995. Information revolution. National Geographic 188 (4):5–37.

USDA–NRCS. 1995. Chief's forum, "Is there a Better Way?" Final report data (Part II). Washington D.C.

Warkentin, B.A. 1994. The discipline of soil science: How should it be organized? Soil Sci. Soc. Am. J. 58:267–268.

Wesley, R.A., and L.A. Smith. 1991. Response of soybeans to deep tillage with controlled traffic on clay soils. Trans. ASAE 34:113–119.

Index